FLOWERING PLANTS OF SEYCHELLES

FLOWERING PLANTS
OF SEYCHELLES

(An annotated Check List, of Angiosperms and Gymnosperms, with Line Drawings)

S.A. ROBERTSON

ROYAL BOTANIC GARDENS
KEW

First published 1989

General editor M.J.E. Coode. Special editor for this volume R.M. Polhill. Cover design, depicting *Lodoicea maldivica* and *Nepenthes pervillei*, by Rosemary Wise; figure on title page of *Medusagyne oppositifolia* from an original drawing by W.E. Trevithick. Text set by Pam Arnold, Christine Beard, Brenda Carey, Margaret Newman, Helen O'Brien and Pamela Rosen.

ISBN 0 947643 14 1

PREFACE

Small islands are most sensitive to occupation by man and require careful development if severe degradation of the environment and inadequate realization of the agricultural potential are to be avoided.

The continuing prosperity of the people of Seychelles depends considerably upon the wise exploitation of their plant resources and as a foundation for this a comprehensive and accurate list of species is of great value. Past works on the flora of Seychelles are now out of date and a book such as the one offered here is welcomed.

Ann Robertson spent six years in Seychelles, accompanying her entomologist husband on his mission against the coconut beetle. Soon after her arrival she realized that a modern floristic work on this region was urgently needed. Being a wise woman she recognized the fact that a flora in the classical style was not within her means but a check-list, which would have much value, was the type of work which she was equipped both with time and ability to produce.

During the period of her residence she sought to catalogue all the different plants growing in the region and to record them from as many different islands as possible. She has not restricted herself to the indigenous species but has included crop plants, weeds and ornamentals. This is most valuable as it is these plants, introduced by man, which are now the most important economic element in the plant life of Seychelles; they far outnumber the indigenous species and occupy the most extensive areas on the inhabited islands. Approximately 1140 species are listed, of which three quarters have been introduced by man.

This work should prove both useful to students of the islands' plant resources and interesting to the curious visitor. It will provide a valuable interim list until the Flora of Seychelles, now being written at Paris, is completed. I hope that this book will make everyone aware of the important role which plants play in the wealth of Seychelles.

S.A. Renvoize
Kew, 1982

NOTE — Various plans to publish this book were not effected until it was offered to Kew in March 1988. To expedite its availability to the public it is published as presented in 1982, but a short list of additions and corrections is given at the end. The additional names are included in the indexes.

INTRODUCTION

When I arrived in Seychelles in September 1975, after several years botanising in eastern Africa, I found that a considerable portion of the flora was new to me and was therefore delighted to find there was a small herbarium in the Ministry of Agriculture. In due course I discovered that, although Jeffrey and Procter had formed a good basic collection, some of the commonest plants were not represented, and, with the permission of the Principal Secretary of the Ministry, I began to add material. I was also able to visit several islands not previously well collected and was therefore able to add to the data on distribution contained in the herbarium. I prepared a catalogue of the plants in the collection and this formed the basis for an updated checklist, Summerhayes' of 1931 being the most recent for the granitic islands. When I knew, in mid 1981, that I would be leaving Seychelles at the end of the year, I began to collect together all the information I had gathered, and with the encouragement of the authorities there, to put it together as a small book which will, I hope, be of use to anyone interested in the plants of Seychelles. Fortunately I was able to spend three months at Kew in early 1982 and completed the book while settling into a new home in Kenya. The list contains 1139 species, in 669 genera divided among 155 families, and only about 250 are thought to be indigenous or endemic.

The checklist is arranged by families following the modified Bentham and Hooker system in use at Kew, and Kew numbers are indicated in brackets after each family name. Within each family the genera and species are arranged alphabetically, and there is an index of family and generic names, and a separate index of vernacular names.

For each species I have shown the source of the name, either from published material (see list of source material), or from herbarium sheets in Seychelles or at Kew. I was unable to visit any other herbaria. I have quoted all the specimens in the Seychelles herbarium and where duplicates occur in Seychelles and at Kew, I have quoted the Seychelles specimens only. This list therefore provides an index to the Seychelles herbarium to November 1981. Kew specimens are quoted only if they do not appear in the source material, or if there is doubt about the name. No specimens have been quoted for the Aldabra group as Fosberg (2) is a new flora and little of the material he used was available to me. A list of collectors is given.

I have used Summerhayes as the starting point for the list and have not consulted the references he quoted, although I have looked at the specimens he refers to which are at Kew. I have quoted those specimens only if material has subsequently been split into different species. As most of the specimens quoted by Summerhayes were collected prior to 1910 it is possible that any species which has only a Summerhayes reference may no longer exist.

I have put vernacular names as given by Bailey, Procter (1), Fosberg (2) and other source material, and as given on herbarium sheets, and have not attempted to rationalize the spelling. It should be noted that Bailey's 'Creole' is a mixture of Creole and French and that his may differ slightly from the Creole names recorded on Aldabra by Fosberg. I have also given English names where these are in common use in Seychelles or in other parts of the tropics.

Where two or more names occur in the source material for the same species I have used the most recent name and noted the other(s). Names in brackets are not necessarily synonyms, but often misapplied names. Some older names have been updated. Some of the names given by Bailey are mis-spelt and in general have incorrect author references and I have righted these errors without comment. I have tried to be accurate with the conventional abbreviations for authors, but am aware that I have probably made some errors. In linking Summerhayes' and Fosberg's (2) lists there are some discrepancies in the treatment of certain genera and I have used my own discretion on this. I have not given author references for the synonyms in the interests of brevity as this will be dealt with in a full Flora.

The checklist covers all the islands in the Seychelles, granitic and coralline, and therefore includes the unique flora of the Aldabra group. I have given localities by island for each species as noted in the source material or on herbarium sheets. Summerhayes' 'Horne' comments have not been included. Bailey's list was included as source material as I have assumed that if he gives a name then that plant must have been seen in Seychelles, but he gives no locality data. If therefore there is no locality recorded for a species, and I have myself seen it, but not collected it, I have noted it as a sight record only (sr only). This list of localities should not be taken as a definitive distribution list as several of the islands have not been collected systematially, nor have plant lists been compiled. A list of the abbreviations of island names referred to in the text is given, together with a map.

I have written a short description of each species, in simple terms, from my own observations of live plants or herbarium material, or from the reference books listed. Again I have felt that a full Flora will provide fuller and more accurate botanical descriptions and keys. Species noted in the source material as endemic or indigenous are so described and I have noted for other plants whether they are pantropical weeds, are cultivated, or are ornamentals. Baumer has been used as a source to note which species have medicinal uses in Seychelles.

I have attempted to illustrate a small proportion of species with simple line drawings and where possible these were drawn from life.

I have included all ornamental and cultivated plants mentioned in the source material, mainly Bailey. Although this means the inclusion of such plants as 'cabbage' which are unlikely to be met with outside a tended garden, it has enabled me to avoid any subjective bias towards those plants which I might consider not part of the flora of Seychelles. The climate of the islands is so favourable to plant growth that plants which must have been spread by man flourish long after all other signs of him have vanished. Thus a seed dropped from a fruit consumed while walking in the forest may germinate as an exotic tree, and ornamentals planted in a long since abandoned garden on a coral island may still thrive.

This list is meant to be only an interim aid to those people interested in plants in Seychelles. In due course a complete Flora is to be produced and Dr Freidman from ORSTOM, Paris is at present on the islands collecting and working on material which will be used to produce this Flora. My list is inevitably incomplete as I did not have the time or resources to collect as widely as I should have liked, or to visit other herbaria where there are Seychelles specimens. I must emphasize that although I had a great deal of help and encouragement from many botanists any errors and omissions are entirely my responsibility.

I am indebted to the following;

The Principal Secretary, Ministry of Agriculture, Seychelles, and many of his staff, for the use of the herbarium and for friendly assistance at all times, and for putting up with an 'unofficial' botanist.

The owners and management of Darros, Poivre, North, Fregate, Coetivy, Platte,

Remire, Silhouette, Alphonse and Therese islands, and the owners of many estates, for allowing me to collect and in many cases for their generous hospitality.

The Director of the Royal Botanic Gardens, Kew, and the Keeper of the Herbarium and his staff, particularly Mr S.A.Renvoize, for identifying many specimens over the years, for permitting me to work at Kew and use the collections and the library, and for encouragement to produce this checklist.

The Botanist in Charge, Miss C.Kabuye, and her staff, of the The Herbarium, National Museums, Kenya, for identifications and also much encouragement.

Dr D.Stoddart, Department of Geography, Cambridge, for help and cooperation.

Mr G.Lionnet and many other friends in Seychelles over the years for assistance in obtaining specimens, names, uses, etc including Mr C.Adam, Mrs E.Beckett, Mr L.Chong Seng, Mr and Mrs C.Lomer, Mr M.Mason, Dr M.Nicoll, Mr D.Todd, Mr S.Warman and Dr J.Watson.

The Government of Seychelles and the British High Commission, for both of whom I worked for varying periods in a clerical capacity, for being understanding about my botanical interests.

And my husband, Ian Robertson, for his constant encouragement, assistance and financial support throughout our stay in Seychelles which enabled me to collect on outlying islands as well as on Mahe and to turn our house into a herbarium cum botanical garden cum office. It is appropriate also to thank his employers, the Overseas Development Administration, UK, for providing the opportunity for me to be in Seychelles for six years as wife of the Team Leader of the Coconut Pest Project, and, fortuitously, for providing me with the chance to work at Kew for three months while we were waiting in UK to be assigned to a project in Kenya.

Waa, Kenya
July 1982

Source Material

The following is a list of source material, arranged alphabetically by author, used for the list of species and for localities. The main sources are italicised. Others were used only if they added to the number of species, or extended the list of localities, not already obtained from the main sources and from herbarium sheets. A few were consulted, but not used, and they are included in this list for completeness. The first column shows the authority as used in the text.

Bailey
   Bailey,D. List of the flowering plants and ferns of Seychelles and their vernacular names. 3rd. ed. Government Printer, Seychelles. 1971

Bailey (not used)
   Bailey,L.H. Palmae Seychellarum. Gentes Herb. 6:1–48. 1942

Baumer
   Baumer,M. Compendium des plantes medicinales des Comores, des Mascareignes et des Seychelles. Fas. 2. Agence de Cooperation Culturelle et Technique. Montpellier. 1979

Evans
   Evans,P.G.H. et al. Preliminary Report, Aberdeen University Expedition to Praslin, Seychelles. Aberdeen. 1976

Fosberg (1)
    Fosberg,F.R. Cousin systematic list of plants. Typescript. Ministry of Agriculture, Seychelles. 1970

Fosberg (2)
    Fosberg,F.R. & Renvoize,S.A. The Flora of Aldabra and neighbouring islands. Kew Bulletin Additional Series VII. HMSO. 1980

Gwynne
    Gwynne,M.D. & Wood,D. Plants collected on islands in the western Indian Ocean during a cruise of the MFRV 'Manihine', Sep–Oct 1967. Atoll Research Bulletin, 134. Smithsonian Institution. 1969 (not used for Aldabra group)

Jeffrey
    Jeffrey,C. The botany of Seychelles: report by the visiting botanist of the Seychelles Botanical Survey, 1961–62. Kew, London. 1962

Lionnet (1)
    Lionnet,G. Striking Plants of Seychelles. Seychelles Nature Handbook No. 5. Ministry of Agriculture, Seychelles. c 1975

Lionnet (2),(2a)
    Lionnet,G. Proposed Guide to the Victoria Botanic Gardens. Cyclostyled pamphlet, Ministry of Agriculture, Seychelles. c 1965 (and draft revision, personal communication (2a))

Piggott
    Piggott,C.J. A report on a visit to the outer islands of Seychelles between October and November 1960. Directorate of Overseas Surveys, London. 1969

Procter (1)
    Procter,J. The endemic flowering plants of the Seychelles: an annotated check list. Candollea 29:345–387. 1974

Procter (2)
    Procter,J. Vallee de Mai Nature Trail Notes. Ministry of Agriculture, Seychelles. c 1972

Procter (3)
    Procter,J. A checklist of the native and naturalized orchids of the Seychelles. Cyclostyled pamphlet, Ministry of Agriculture, Seychelles. 1973

Renvoize
    Renvoize,S.A. A floristic analysis of the western Indian Ocean coral islands. Kew Bulletin 30(1):133–152. 1975

Robertson (1)
    Robertson,S.A. & Fosberg,F.R. A list of plants collected on Platte Island. Atoll Research Bulletin, 273. 1983

Robertson (2)
    Robertson,S.A. & Fosberg,F.R. Plants of Coetivy. Atoll Research Bulletin, 273. 1983

Robertson (3)
    Robertson,S.A. & Fosberg,F.R. List of plants of Poivre Island, Amirantes. Atoll Research Bulletin, 273. 1983

Robertson (4)
Robertson,S.A. & Todd,D.M. Vegetation of Fregate Island, Seychelles. Atoll Research Bulletin, 273. 1983

Robertson (5)
Robertson,S.A. A list of plants collected on North Island. (in preparation)

Robertson (6)
Robertson,S.A. A list of plants seen on St Anne Island. Typescript, Ministry of Agriculture, Seychelles, 1981

Robertson (7)
Robertson,S.A. A list of plants of D'Arros Island. Typescript with the plant collection, Darros, Seychelles. 1980

Robertson (8)
Robertson,S.A., Robertson,I.A.D. & Fosberg,F.R. A list of plants collected on Alphonse Island, Seychelles. Atoll Research Bulletin, 273. 1983

Sauer
Sauer,J.D. Plants and Man on the Seychelles coast. University of Wisconsin Press. 1967

Stoddart (1)
Stoddart,D.R. Coral islands of the western Indian Ocean, Atoll Research Bulletin, 136. Smithsonian Institution. 1970 (not used for Aldabra group)

Stoddart (2)
Stoddart,D.R., Coe,M.J. & Fosberg,F.R. D'Arros and St Joseph, Amirante Islands. Atoll Research Bulletin, 223. Smithsonian Institution. 1979

Stoddart (3)
Stoddart,D.R. & Fosberg,F.R. Bird and Denis Islands, Seychelles. Atoll Research Bulletin, 252. Smithsonian Institution. 1981

Summerhayes
Summerhayes,V.S. An enumeration of the angiosperms of the Seychelles archipelago. Trans. Linn. Soc. (Zool.) Ser. 2. Vol. 19. pt. 2. 1931

Sumner Gibson
Sumner Gibson,H.M. Report on the forests of Seychelles. Government Printer, Seychelles. 1938

Swabey
Swabey,C. Forestry in the Seychelles (Report of a visit by the Forestry Adviser in 1959). Government Printer, Seychelles. 1960

Vesey Fitzgerald (not used)
Vasey Fitzgerald,D. On the vegetation of Seychelles. J. Ecology 28(2):465–483. 1939

Vesey Fitzgerald (not used)
Vesey Fitzgerald,D. Further studies of the vegetation on islands in the Indian Ocean. J. Ecology 30(1):1–16. 1942

Wilson (1)
Wilson,J.R. Preliminary plant list, Vallee de Mai, Praslin. Cyclostyled list, Ministry of Agriculture, Seychelles. 1980

Wilson (2)
>   Wilson,J.R. Preliminary plant list, Curieuse National Park. Cyclostyled list, Ministry of Agriculture, Seychelles. 1980

Wilson (3)
>   Wilson,J.R. Ecology of Desnoeufs, Amirantes Islands. Atoll Research Bulletin, 273. 1983

Wilson (4)
>   Wilson,J.R. Ecology of Marie Louise, Amirantes Islands. Atoll Research Bulletin)

References consulted

Agnew,A.D.Q. Upland Kenya Wild Flowers. Oxford University Press. 1974
Bailey,L.H. & Bailey,E.Z. Hortus Third. Macmillan. 1976
Bruggeman,L. Tropical Plants and their Cultivation. Thames and Hudson. 1962
Bosser,J. et al. Flore des Mascareignes. MSIRI, ORSTOM et Kew (ODA). (1976–present, in progress)
Dale,I.R. & Greenway,P.J. Kenya Trees and Shrubs. London and Nairobi. 1961
Everard,B. Wild Flowers of the World. Rainbird Reference Books Ltd. 1970
Graf,A.B. Tropica. Roehrs Co. USA. 1978
Heywood,V.H. ed. Flowering plants of the World. Oxford University Press. 1978
Hubbard,C.E., Milne-Redhead,E., & Polhill,R.M. eds. Flora of Tropical East Africa. Crown Agents, London. (1952–present, in progress)
Ivens,G.W. East African Weeds and their Control. Oxford University Press. 1967
Jex-Blake,A.J. Gardening in East Africa. Longmans. 1950
Kokwaro,J.O. Medicinal Plants of East Africa. East African Literature Bureau. 1976
Lind,E.M. & Tallantire,A.C. Some Common Flowering Plants of Uganda. Oxford University Press. 1962
MacMillan,H.F. Tropical Planting and Gardening. Macmillan. 1943
Merrill,E.D. A Flora of Manila. Manila Bureau of Printing. 1912
Napper,D.M. Cyperaceae of East Africa. Journal of the East African Natural History Society 106, 109, 110, 113, 124. Nairobi. 1963–71
Napper,D.M. Grasses of Tanganyika. Ministry of Agriculture, Forests and Wildlife, Bulletin 18, Tanzania. 1965
Purseglove,J.W. Tropical Crops. Longmans. 1968
Williams,R.O. Useful and Ornamental Plants of Zanzibar and Pemba. Zanzibar. 1949
Willis,J.C. Dictionary of the Flowering Plants and Ferns. 6th and 7th editions. Cambridge University Press. 1931 and 1966

Collectors

The following abbreviations are used in the text for collectors who deposited herbarium specimens in Seychelles (S) and/or at Kew (K). I have seen all the specimens quoted. There are a few 'singletons' and these collectors have been named in full in the text. In the few cases where specimens quoted in Summerhayes have had to be split into different species I have quoted his collectors and their numbers.

Arc   Archer,P.G. Collected on Mahe and Praslin in 1960 (K)

Ber   Bernardi,L. Collected on Mahe in 1973 (K)
Bru/War   Brunet,N. & Warman,S. Collected on Aride in 1975 (S) and (K)
Dup   Dupont,P.R. Collected on Mahe 1930-40 (K)
Fos   Fosberg,F.R. Collected on Cousin, Cousine, Mahe in 1970 (S) and (K) and
   also on coral islands and Aldabra group.
Fra   Frazier,J. Collected on Remire 1973 (S)
Fra/Fea   Frazier,J. & Feare,C. Collected on Isle aux Vauches Marine near
   Mahe in 1973 (S)
Gwy/Woo   Gwynne,M.D. & Wood, D. Collected on coral islands in 1967 (K)
Jef   Jeffrey,C. Collected widely 1961–62 (S) and (K)
Lam   Lambert,M. Collected on Mahe in 1970 (K)
Pig   Piggott,C.J. Collected on coral islands, unnumbered specimens as sn only,
   in 1960 (K)
Pro   Procter,J. Collected widely 1970–72 (S) and (K)
Rob   Robertson,S.A. Collected widely 1975–81 (S) and (K). Note: sr = sight
   record (see Introduction)
Rob/Wil   Robertson,S.A. & Wilson,J.R. Collected in 1979–80 (S)
Rob/Gre   Robertson,S.A. & Greig Smith, J. Collected on Silhouette in 1979 (S)
Sch   Schlieben,H. Collected on Mahe and Praslin in 1970 (K)
Squ   Squibbs,F.L. Collected on Long in 1936 (K)
Sto   Stoddart,D.R. Collected widely on coral islands 1970–80 (K)
Sto/Poo   Stoddard,D.R. & Poore,M.E.D. Collected on coral islands 1968 (K)
Tod   Todd,D.M. Collected on Aride and Fregate in 1981 (S)
Ves   Vesey-Fitzgerald,D. Collected widely in 1938–39 (K)
War   Warman,S. Collected on Aride in 1978 (S)

Localities — see list and maps overleaf.

Localities

The following abbreviations have been used in the text, in the order indicated, granitic islands followed by coral islands.

Granitic islands (in order of size)	Coral islands (roughly from north to south)
M Mahe (BG) Botanic Gardens	Bir Bird
S Silhouette	Den Denis
P Praslin	Pla Platte
D La Digue	Coe Coetivy
F Fregate	Afr African Banks
N North	Rem Remire
C Curieuse	Des Desroches
A Aride	Dar Darros
Fe Felicite	Jos St Joseph Atoll
Ma Marianne	Poi Poivre
An St Anne	Mar Marie Louise
Cn Conception	Def Desnoeufs
T Therese	Alp Alphonse
Ce Cerf	Pro Providence
L Long	Pie St Pierre
Co Cousin	Far Farquahar
Cu Cousine	Ald Aldabra
Mo Moyenne	Ass Assumption
R Recif	Cos Cosmoledo
V Isle aux Vache Marine (Mahe)	Ast Astove
Se Isle Seche (= Beacon)	
Ro Round	
B Beacon	
Ao Isle Anonyme	

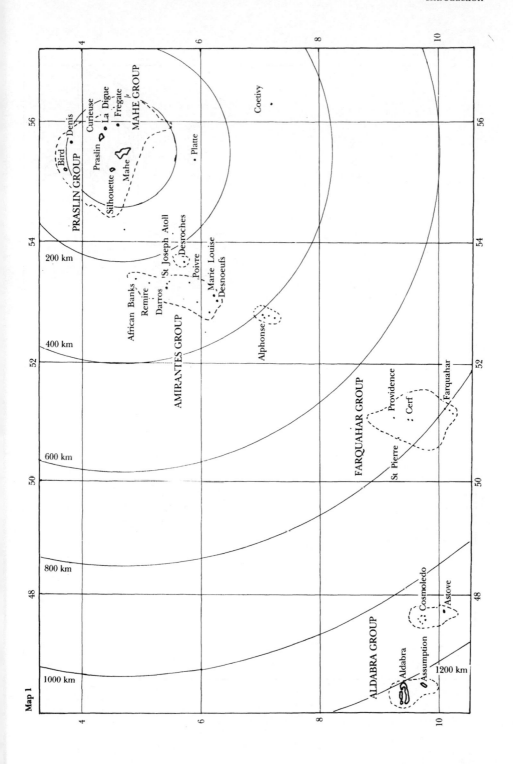

Map 1

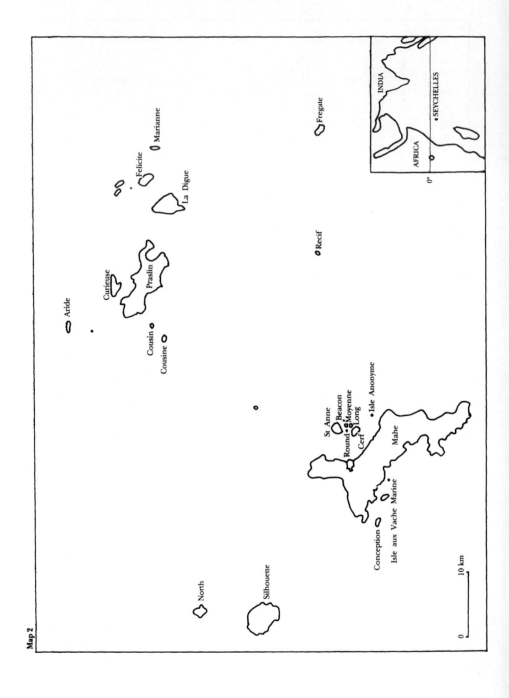

Map 2

THE CHECK LIST

D I C O T Y L E D O N S

RANUNCULACEAE (1)

Delphinium L.

D. cardiopetalum L. 'pied d'alouette'

Bailey

No locality recorded

Cultivated herb with deeply lobed leaves to 20 cm long and spikes to 2 m of spurred blue, pink or white flowers. The nomenclature of cultivated species is confused.

DILLENIACEAE (2/1)

Dillenia L.

D. ferruginea (Baillon)Gilg (*Neowormia ferruginea*) 'bois rouge'

Summerhayes, Bailey, Procter (1), (2)

Jef 450(S),640(K), Sch 11736(K), Ber 14628(K), Rob 2535(S)

Localities recorded: M,S,P,C

Endemic tree with ovate leaves to 40 cm long, rusty tomentose and strongly ribbed below, often holed by insects, white flowers to 4 cm wide and orange fruit to 1 cm wide hidden in the sepals. The young plants have larger, stem-clasping leaves.

D. indica L. 'elephant apple, gaufrier, grain mon pere'

Bailey, Lionnet (2)

Jef 587(S)

Localities recorded: M(BG)

Cultivated tree to 20 m with ovate, serrated edged leaves to 35 cm long, white flowers to 20 cm wide and fruit within the enlarged sepals, globose, to 15 cm wide and smelling of burnt rubber.

D. suffruticosa (Griff.)Martelli 'bois rouge blanc'

Bailey

Jef 1158(S), Pro 3982(K)

Localities recorded: M

Cultivated and escaping shrub with ovate leaves to 40 cm long or more, yellow flowers to 10 cm wide with white stamens and fruit capsules which open star-fashion to show pink inside and orange brown seeds. (Fig. 1)

MAGNOLIACEAE (4/1)

Michelia L.

M. champaca L.   'champac'

Bailey

No locality recorded

Cultivated tree to 20 m with long pointed ovate leaves to 25 cm long, yellow or orange fragrant flowers to 6 cm wide and fruiting spikes to 20 cm long with fleshy ovoid spotted fruits containing seeds in a pinky grey pulp.

ANNONACEAE (5/1)

Annona L.

A. cherimola Mill.   'cherimolia, cherimoya'

Bailey, Robertson (5)

Localities recorded: M(sr only), N

Cultivated tree to 8 m with ovate leaves to 15 cm long, velvety brown beneath, yellowish fragrant flowers to 2 cm long and edible heart-shaped green fleshy fruits to 12 cm long.

A. diversifolia Saff.   'corossol de france, ilama'

Bailey

Localities recorded: M(sr only)

Cultivated tree to 8 m with elliptic leaves to 12 cm long, maroon flowers to 2.5 cm long and pinkish edible globose fleshy fruit to 15 cm long with tubercles.

A. muricata L.   'corossol, soursop'

Bailey, Fosberg, Baumer, Wilson (1), Robertson (4), (5)

Localities recorded: P,F,N,Co, M(sr only)

Cultivated tree to 6 m with dark green ovate leaves to 15 cm long, yellow fragrant flowers to 2.5 cm long and dark green ovoid edible fruits to 25 cm long, covered with recurved fleshy spines. Used medicinally in Seychelles.

A. reticulata L.   'coeur de boeuf, bullock's heart'

Bailey, Robertson (4),(5)

Localities recorded: M(sr only), F,N

Cultivated tree to 12 m with lanceolate leaves to 18 cm long, yellowish flowers to 2.5 cm long and brownish red edible, rather warty fruit to 10 cm wide.

A. squamosa L.   'zatte, attier, custard apple, sweetsop'

Bailey, Fosberg & Renvoize, Robertson (2),(3),(4),(5)

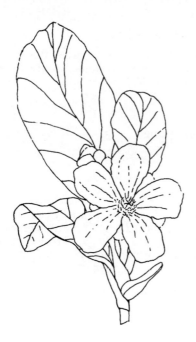

Fig. 1. Dillenia suffruticosa

Fig. 2. Annona squamosa

Fig. 3. Argemone mexicana

Fig. 4. Gynandropsis gynandra

3

Fos 52123(S), Bru/War 50(S)

Localities recorded: M(sr only), F,N,A,Co,Coe,Poi,Ald

Cultivated tree to 7 m with ovate leaves to 10 cm, yellowish green flowers to 2 cm long and heart-shaped edible fruit to 10 cm wide, made up of rounded pale green fleshy scales. (Fig.2)

Cananga (DC.)Hook.f. & Thoms.

C. odorata (Lam.)Hook.f. & Thoms.   'ylang ylang'

Bailey, Wilson (1), Robertson (4)

Jef 756(S)

Localities recorded: M,P,F

Cultivated and escaping tree to 25 m with drooping branches and ovate pointed leaves to 20 cm long, pendent long petalled heavily scented greenish yellow flowers to 7.5 cm long and black fleshy unevenly ovoid fruit to 2.5 cm long. The flowers yield an essential oil used for perfumery.

MENISPERMACEAE (6)

Cissampelos L.

C. pareira L.   'la liane, ice vine'

Renvoize, Fosberg (2)

Localities recorded: Ass

Introduced weed, a climber with peltate leaves to 10 cm wide, minute yellowish flowers and small scarlet globose fruit to 5 mm wide.

NYMPHAEACEAE (8/1)

Nymphaea L.

N. caerulea Savigny   'blue lotus, water lily'

Lionnet (2)

Jef 594(S)

Localities recorded: M(BG)

Cultivated and escaping aquatic plant with heart-shaped floating leaves to 35 cm wide and flowers with spotted sepals and blue petals to 6 cm long.

N. nouchali Burm.f. (*N.stellata*)   'lis d'eau, nenuphar etoile, tam tam'

Bailey, Wilson (1)

Localities recorded: P

Cultivated and escaping aquatic plant with heart-shaped and sinuate-edged leaves to 20 cm wide and flowers to 12 cm wide with pink outer and blue inner petals.

PAPAVERACEAE (10)

Argemone L.

A. mexicana L.   'chardon du pays, mexican poppy'

Bailey

Rob 3140(S)

Recorded localities: Coe

Pantropical weed, a herb to 1m with grey green spiky lobed leaves to 15 cm long, yellow flowers to 6 cm wide and fruit an oblong valved capsule to 5 cm long containing many small seeds. (Fig. 3)

CRUCIFERAE (*BRASSICACEAE*) (11)

Brassica L.

B. chinensis L.   'bred chou de chine, pe-tsai, chinese cabbage'

Bailey, Robertson (4)

Recorded localities: M(sr only), F

Cultivated leafy vegetable, the leaves used as spinach or for salad.

B. nigra (L.) Koch (*B.juncea*)   'pissard de chien, black mustard'

Fosberg (2)

Recorded localities: Ald,Ass,Cos,Ast

Cultivated herb with yellow flowers and erect narrowly oblong fruits to 5 cm long containing many spherical brown seeds which are used as a spice.

B. oleracea L. var. **bullata**   'chou, cabbage'

Bailey

Localities recorded: M(sr only)

Cultivated vegetable with tightly packed leaves forming a head.

B. oleracea L. var. **botrytis**   'chou-fleur, cauliflower'

Bailey

Localities recorded: M(sr only)

Cultivated vegetable with leaves to 50 cm and a terminal white inflorescence of tightly packed flowers which is eaten before the flowers open.

B. oleracea L. var. **gongyloides** (*B.caulo-rapa*)   'chou navet, kohl-rabi'

Bailey

Localities recorded: M(sr only)

Cultivated vegetable with swollen white edible stem bearing a few leaves and a tuft of leaves at the top.

Nasturtium R.Br.

N. officinale R.Br. (*Roripa aquaticum*)   'cresson, watercress'

Bailey, Robertson (4)

Localities recorded: M(sr only), F

Cultivated vegetable with creeping stems and pinnate leaves to 15 cm long which are used as salad.

Raphanus L.

R. sativus L.   'radis, radish'

Bailey

Localities recorded: M(sr only)

Cultivated vegetable of which the swollen red or white root is eaten.

CAPPARACEAE (12/1)

Capparis L.

C. cartilaginea Decne. (*C.galeata*)   'bois zanguette'

Bailey, Renvoize, Fosberg (2)

Localities recorded: Ald,Ass,Ast

Indigenous spreading armed shrub with tomentose grey branches, widely obovate leaves notched at the apex to 5 cm long, white flowers with many 5 cm long stamens and red ovoid ribbed fruit.

Cleome L.

C. strigosa (Boj.)Oliv.   'bred caya'

Renvoize, Fosberg (2)

Localities recorded: Ald,Ass,Cos,Ast

Annual herb with hairy 3–5-foliolate leaves, the leaflets obovate and the longest to 3 cm long, deep purple flowers and erect slightly club-shaped capsules splitting from the base when ripe.

C. viscosa L.   'pissat de chine'

Summerhayes, Bailey, Gwynne, Stoddart (1),(3), Robertson (6), Baumer, Fosberg (1), Renvoize

Jef 656(S), Pro 4054(S), Rob 2411(S), Tod 20(S)

Localities recorded: M,S,F,N,Fe,An,Co,Bir,Coe,Rem

Pantropical weed, sticky, with 3 to 5 foliolate leaves, the leaflets obovate to 5cm long, yellow flowers and an erect striated capsule to 7cm long. Used medicinally.

Gynandropsis DC.

G. gynandra (L.)Briq.   *(Cleome gynandra, Gynandropsis pentaphylla)*

Summerhayes, Gwynne, Renvoize, Fosberg (2), Stoddart (1),(3), Wilson (3),(4), Robertson (3)

Jef 1173(S), Pro 4046(S), Rob 2313(S)

Localities recorded: P,F,Bir,Rem,Poi,Des,Mar,Def,Pie,Far,Ald

Pantropical weed, pubescent, with 5-foliolate leaves, the leaflets rhomboidal to 6cm long, white flowers tinged with purple and stamens on a gynophore, and a narrow flattened capsule to 7 cm long. (Fig. 4)

Maerua Forssk.

M. triphylla A.Rich.   'bois trois feuilles'

Renvoize, Fosberg (2)

Localities recorded: Ald, Ass, Cos, Ast

Small tree or shrub with 3–4-foliolate leaves, many small cream flowers and cylindrical woody fruits constricted between the seeds.

VIOLACEAE (15)

Viola L.

V. odorata L.   'violette, sweet violet'

Bailey

No locality recorded

Cultivated stoloniferous bedding plant with heart-shaped leaves to 10cm wide and fragrant violet or white spurred flowers to 2 cm wide.

BIXACEAE (17/1)

Bixa L.

B. orellana L.   'roucou, annatto'

Summerhayes, Bailey, Baumer

Localities recorded: M(sr only)

Cultivated shrub to 12 m with ovate or heart-shaped leaves to 25 cm long, pale pink or white flowers to 6 cm wide and brown burred capsules to 4 cm long which open to show the seeds covered with red powder. There is a variety with larger flowers and red fruits. The red powder on the seeds is used as a food colouring. The plant is used medicinally in Seychelles. (Fig. 5)

FLACOURTIACEAE (17/3)

Aphloia Ben.

A. seychellensis Hemsley   'bois merle'

Summerhayes, Bailey, Procter, Baumer

Ves 5502(K), Jef 545,779(K),726(S), Sch 11762,11791(K), Fos 51978(S), Pro 4139(S),4531(K)

Localities recorded: M,S,P

Endemic small forest tree with stiff ovate finely serrated edged leaves to 10 cm long, small white flowers to 1 cm wide and globose white fruits to 1 cm wide containing many seeds. Used medicinally in Seychelles.

A. sp.   'bois mare petite feuilles'

Bailey

No locality recorded.

Flacourtia Comm. ex L'Hér.

F. ramontchi L'Hér. var. **ramontchi** (*F.indica,F.cataphracta*)   'prune, prune marron, madagascar or indian plum'

Summerhayes, Bailey, Baumer, Wilson (1)

Pro 4008(S),4535(K), Rob 2409(S)

Localities recorded: M,S,P,?Pla

Tree to 3 m with branched spines on the trunk, pointed ovate leaves to 10 cm long with toothed edges, small flowers and purple edible ovoid juicy fruit to 3 cm long with a few flattened seeds. There are ♂ and ♀ trees and it is used medicinally in Seychelles.

F. ramontchi L'Hér. var. **renvoizei** Fosberg   'prunier, prune'

Renvoize, Fosberg (2)

Localities recorded: Ald, Ast

Endemic small unarmed tree with ovate leaves to 7 cm long and ovoid black fruit to 1 cm long in clusters of up to 8.

Hydnocarpus Gaertn.

H. pentandra (Buch.-Ham.)Oken   (*H.anthelmintica, H.wightiana, H.laurifolia*) 'choulmoogra'

Bailey, Swabey

Pro 4495(K) Rob 2390(S)

Localities recorded: M,C

Cultivated tree to 15 m with lanceolate leaves to 20 cm long, small white flowers and large rusty brown warty irregularly globose fruit to 15 cm long containing a few large seeds. The oil was used in the past to treat leprosy.

Fig. 5. Bixa orellana

Fig. 6. Ludia mauritiana

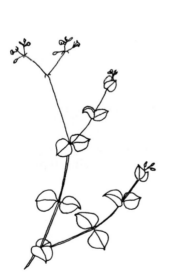

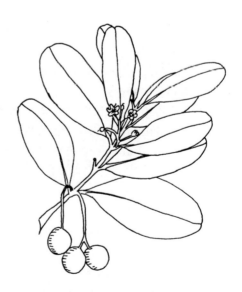

Fig. 7. Drymaria cordata

Fig. 8. Calophyllum inophyllum

Ludia Comm. ex Juss.

L. mauritiana Gmel. (*L.sessiliflora*)   'prunier marron, petit prun'

Summerhayes, Fosberg (2)

Ves 5435(K), Pro 4007,4058,4258(S), Rob 2769(S)

Localities recorded: M,S,P,C,Fe,Ald

Indigenous small tree with erect branches dotted with pale lenticels, thick leathery net veined ovate or lanceolate leaves to 12 cm long, small flowers with no petals but many stamens sessile in the leaf axils and beaked globose red fruit to 1 cm long. (Fig. 6)

PITTOSPORACEAE (18)

Pittosporum Banks

P. wrightii Hemsley   'bois joli coeur'

Summerhayes, Bailey, Procter (1), Baumer

Ves 6194(K), Jef 809,1216(S), Pro 4246(S),4004(K), Rob 2539(S)

Localities recorded: M,S

Endemic shrub with greyish branches, pointed obovate leaves to 12 cm long which smell of carrots, white flowers to 7 mm wide and orange globose fruits to 8 mm long splitting to show 4 sticky red seeds. Used medicinally in Seychelles.

CARYOPHYLLACEAE (22/1)

Drymaria Willd.

D. cordata (L.) Roem. & Schultes

Summerhayes

Jef 438(S), Fos 51975(S), Rob 2530,2998,2613(S), Tod 14(S)

Localities recorded: M,F,N,Pla

Pantropical weed, a slender straggling herb with opposite round leaves to 1.5cm wide, small white flowers and sticky flower stalks, sepals and fruit. (Fig. 7)

Dianthus L.

D. chinensis L.   'oeillet de chine, chinese or indian pinks'

Bailey

Localities recorded: M(sr only)

Cultivated garden annual with opposite linear leaves clasping the stem to 7 cm long and few–many-flowered inflorescence of flowers to 2.5 cm wide, pinkish with darker centres.

PORTULACACEAE (23/1)

Portulaca L.

P. grandiflora L.   'gorge tortue, portulaca'

Bailey

No locality recorded

Garden ornamental, a fleshy spreading herb with sausage-shaped leaves to 2.5 cm long and gaudy single or double flowers, usually purple, to 3 cm wide.

P. mauritiensis Poelln. (*P.quadrifida* not L., *P.australis*)   'pourpier'

Renvoize, Fosberg (2), Stoddart (1)

Rob 2308(S)

Localities recorded: Afr,Rem,Far,Ald,Ass,Cos,?Ast

Prostrate fleshy herb with ovate leaves to 2.5cm long and deep yellow flowers with the sepals not keeled, the nodal appendages hairy. This is probably an indigenous species with an endemic subspecies on Ald, Ass and Cos.

P. oleracea L.   'pourpier, portulaca, purslane'

Summerhayes, Bailey, Gwynne, Renvoize, Stoddart (1),(2),(3), Fosberg (2), Baumer, Wilson (2),(3),(4), Robertson (3)

Jef 622(S),Fos 52039(S),Bru/War 55(S),Rob 2307,2507,2587,2728,2949,3033,3137(S)

Localities recorded: M,F,N,A,Fe,C,L,Co,Cu,Bir,Den,Pla,Coe,Afr,Rem,Dar,Jos,Poi,Mar,Def,Pie,Far,Ald,Ass,Cos,Ast

Pantropical weed (perhaps an endemic subspecies on Ald), a glabrous prostrate fleshy herb with obovate leaves to 2.5 cm long and yellow flowers, the sepals keeled. The leaves can be eaten as a spinach and the plant is used medicinally in Seychelles.

Talinum Adans.

T. paniculatum (Jacq.)Gaertn.

Rob 2993(K)

Localities recorded: M(sr only),Dar

Cultivated and escaping succulent tuberous rooted herb with basal rosette of slightly fleshy obovate leaves to 7 cm long, slender erect flower stalks with pink flowers and orange capsules.

GUTTIFERAE (*CLUSIACEAE*) (27)

Calophyllum L.

C. inophyllum L.   'takamaka, alexandrian laurel'

Summerhayes, Sumner Gibson, Bailey, Renvoize, Fosberg (2), Gwynne, Lionnet (1), Stoddart (1),(2),(3), Wilson (4), Robertson (3),(4),(5),(6),(7)

Jef 492(S), Fos 52141(S), Sch 11746(K), Bru/War 21(S), Rob 3119(S)

Localities recorded: M,S,P,D,F,N,C,A,Fe,An,Co,Bir,Den,Coe,Rem, Des,Dar,Poi,Mar,Far,Ald,Ass,Cos,Ast

Indigenous evergreen strand tree with opposite obovate leathery leaves to 20 cm long, the lateral veins very fine and parallel, white flowers to 2 cm wide with many stamens and red ovary and round or ovoid green fruit to 2 cm wide. (Fig. 8)

C. parviflorum Boj. ex Bak.   'takamaka madagascar'

Sumner Gibson, Bailey

Localities recorded: M

Tree with opposite ovate leathery leaves to 7 cm long, finely parallel veined, white flowers to 1.5 cm wide and pointed narrowly ovate green fruit to 3 cm long. (Description from Mauritius material). (See *Noronhia emarginata*).

Garcinia L.

G. mangostana L.   'mangoustan pourpre, mangosteen'

Bailey, Lionnet (2), Baumer

Localities recorded: M(BG)

Cultivated small tree with opposite elliptic leathery leaves to 25cm long, solitary pink flowers to 5cm wide, the petals falling early to leave the bright red sepals, and globose purplish fruit bearing the star-shaped stigma and containing edible white jelly around the few seeds. Used medicinally in Seychelles.

G. xanthochymus Hook.f.   'mangoustan jaune (aigre)'

Bailey, Lionnet (2)

Localities recorded: M(BG)

Cultivated tree to 12 m with opposite oblong leathery leaves to 50 cm long and white flowers to 2 cm wide with dark yellow globose fruit to 7 cm wide. Edible but not as delicious as *C. mangostana*.

Pentadesma Sabine

P. butyracea Sabine   'bois beurre, butter nut tree'

Bailey, Swabey

Rob 2523(S)

Localities recorded: M,P

Cultivated and escaping tree with dark glossy opposite leathery leaves to 20 cm long, white many-stamened flowers to 10 cm long, opening at night, and large brown pear shaped fruit to 15 cm long with yellow oil in the flesh.

THEACEAE (28/1)

Camellia L.

C. sinensis (L.)O.Kuntze (*C.thea*) 'thé, tea'

Bailey, Wilson (1)

Pro 4528(K)

Localities recorded: M,P

Cultivated and escaping tree or shrub with shiny elliptic serrated edged leaves to 12 cm long, white many-stamened flowers to 3 cm wide and hard brown 1 to 4-lobed fruits. The young leaves are fermented slightly and dried and chopped to give the commercial 'tea'. (Fig. 9)

MEDUSAGYNACEAE (28/11)

Medusagyne Baker

M. oppositifolia Baker 'bois medus, jellyfish tree'

Summerhayes, Bailey, Procter (1), Lionnet (1)

Sch 11717(K),Pro 3991(S),Ber 14704(K),Rob 2267, 2482(S),Wil/Rob 3163A(S)

Localities recorded: M

Rare endemic tree with fibrous corky bark, opposite leathery obovate leaves to 8 cm long, notched at the tip and sinuate edged, small white many-stamened flowers to 1 cm wide and ovoid grooved capsules to 1 cm long which open like small brown umbrellas when ripe to release the minute flat winged seeds.

DIPTEROCARPACEAE (29/1)

Vateria L.

V. seychellarum Dyer (*Vateriopsis seychellarum*) 'bois de fer'

Summerhayes, Bailey, Procter (1), Lionnet (1)

Ves 6336(K), Jef (S), Sch 11734(K), Pro 4529(K), Ber 14637(K), Rob 2707(S)

Localities recorded: M

Rare endemic tree to 8 m with large ovate pointed leaves to 25 cm long, few-flowered axillary spikes of whitish or cream flowers to 2 cm wide and rusty brown globose or conical fruit to 4 cm long.

MALVACEAE (31/1)

Abutilon Mill.

A. angulatum (Guill. & Perr.) Masters 'mauve, mauve batard'

Renvoize, Fosberg (2)

Localities recorded: Ald,Cos,Ast

Shrub with stems angular in cross-section, cordate leaves to 11 cm long, inflorescences of orange flowers to 4 cm wide, maroon at the centre, and a yellow tomentose capsule to 1.5 cm wide with about 25 mericarps (valves of the fruit).

A. fruticosum Guill. & Perr.

Renvoize, Fosberg (2)

Localities recorded: Ass

Slender much branched shrub with ovate leaves to 5 cm long, yellow flowers to 2 cm wide and capsules to 1 cm wide with up to 8 mericarps.

A. indicum (L.)Sweet (*A. mauritianum*) 'mauve du pays'

Summmerhayes, Bailey, Gwynne, Renvoize, Stoddart (1),(3), Baumer, Wilson (3),(4)

Fos 52068(S), Pro 4156,4232(S), Fra 438(S),Rob 2504, 2578(S), Tod 26(S)

Localities recorded: S,F,N,Co,Bir,Den,Dar,Rem,?Mar,?Def,?Far

Softly woody herb or shrub with triangular or heart-shaped leaves, greyish beneath and sinuate edged, to 10 cm long, yellow flowers to 2 cm wide closing early in the day and capsule, greenish tomentose at first, with up to 20 black mericarps. Used medicinally in Seychelles. (Fig. 10)

A. pannosum (Forst.f.)Schlect. 'mauve, mauve batard'

Fosberg (2)

Localities recorded: Ald,Cos,Ast

Shrub to 3 m with branches circular in cross-section, heart-shaped leaves to 15 cm wide, small orange yellow flowers to 15 mm long and capsules to 1.5 cm wide with up to 20 mericarps. (This may be the same as *A. indicum* as there appears to be some confusion over the nomenclature)

Gossypium L.

G. hirsutum L. (*G. mexicanum*) 'cotonnier, cotton'

Summerhayes, Bailey, Gwynne, Stoddart (1),(2),(3), Wilson (4), Robertson (3),(7)

Pig sn, Fos 52073(S), Pro 4148(S), Rob 2830,3117(S), Tod 39(S)

Localities recorded: Co,Ro,F,Bir,Den,Coe,Des,Dar,Poi,Mar,Alp,Ald,Ass, Cos,Ast

Cultivated and escaping tall woody herb with alternate 3 to 5 lobed leaves to 15 cm wide, funnel shaped yellow flowers with dark centres to 5 cm wide, turning pink with age, and pointed ovate capsules splitting when ripe to expose the black seeds covered with cotton. Cotton was the original cash crop on the islands before coconuts and the plants on the outlying islands are relics of this early cultivation and there is some suggestion that it was a

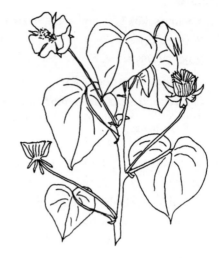

Fig. 9. Camellia sinensis   Fig. 10. Abutilon indicum

Fig. 11. Hibiscus rosa-sinensis   Fig. 12. Hibiscus tiliaceus

'barbadense' type. Recently 'sea island' cotton has been grown experimentally on Mahe.

Hibiscus L.

H. abelmoschus L.   'ambrette, musk okra or mallow'

Fosberg (2)

Rob 2741(S)

Localities recorded: F,Ast

Pantropical stout hairy herb with angular or deeply lobed leaves to 15 cm long, yellow flowers to 8 cm long with red centres and elongated fruit capsules to 8 cm long.

H. esculentus L.   'ambrette, lalo, okra, lady's fingers'

Bailey

Localities recorded: M(sr only)

Cultivated roughly hairy stout herb with 5-lobed leaves to 25 cm long, yellow flowers to 7 cm long and long edible fruit capsules to 30 cm long.

H. liliiflorus Cav.   'augerine, mandrinette, hibiscus'

Summerhayes, Bailey

Localities recorded: M

Shrub with lower leaves deeply and narrowly 3-lobed to 10 cm long, upper leaves obovate to 10 cm long, idented at tip, and white flowers to 10 cm long.

H. mutabilis L.   'passe rose, blushing bride'

Bailey, Robertson (4),(7)

Localities recorded: M(sr only),F,Dar

Cultivated shrub with softly hairy and shallowly 5-lobed leaves to 25 cm long, large white double flowers to 12 cm wide which fade to dark pink.

H. physaloides Guill. & Perr.

Summerhayes

Localities recorded: P

Woody herb with lobed leaves to 5 cm long with sinuate edges, white flowers to 4 cm long with a purple spot at the base of each petal and linear bracts to 8 mm long, and capsules to 1.5 cm long.

H. rosa-sinensis L.   'foulsapate, hibiscus rose, hibiscus'

Summerhayes, Bailey, Wilson (1), Robertson (7)

Jef 449,638(S)

Localities recorded: M,P,Dar

Cultivated shrub to 3 m with glossy ovate sinuate-edged leaves to 10 cm long and large showy variously coloured double or single flowers. (Fig. 11)

H. sabdariffa L. 'roselle, sorrel'

Bailey

No recorded locality

Cultivated woody herb with red stems and deeply lobed leaves to 15 cm long, large red flowers to 5 cm long with a ring of pale yellow on the petals and fruits formed by the sepals swelling to envelop the 3 cm long capsule. These enlarged sepals are a commercial source of dietary Vit C.

H. schizopetalus Hook.f. 'coral hibiscus'

Summerhayes, Robertson (7)

Localities recorded: M,Dar

Cultivated shrub with glossy ovate sinuate-edged leaves to 10 cm long and pendant flowers with incised reflexed petals, white or red, with the staminal column hanging below.

H. surattensis L.

Summerhayes

Rob 3163B(S)

Localities recorded: M,Co

Weakly stemmed herb with prickly stem and deeply lobed leaves to 8 cm long and showy yellow flowers to 5 cm long with purple centres with a ring of leafy bracts followed by prickly pointed capsules to 2 cm long.

H. tiliaceus L. 'varre, mahoe'

Summerhayes, Bailey, Fosberg (2), Renvoize, Lionnet (1), Sauer, Stoddart (2),(3), Wilson (3),(4), Robertson (4)

Jef 411(S), Sch 11744(K), Fos 52170(S), Bru/War 53(S) Rob 2616,3003(S)

Localities recorded: M,F,N,C,A,L,Co,Den,Bir,Pla,Jos,Mar,Def,Ald,Ast

Indigenous shrub with heart-shaped leaves to 20 cm wide, greyish beneath, yellow flowers with dark red centres to 8 cm long, turning red in the afternoon, and globose capsules to 2 cm long, splitting into 5 valves. (Fig. 12)

Malachra L.

M. capitata (L.)L. 'mauve a fleurs jaunes'

Bailey

No locality recorded

Pantropical coarsely hairy herb with shallowly lobed leaves to 20 cm wide and small yellow flowers to 1 cm long in tight axillary clusters.

Malvastrum A. Gray

M. coromandelianum (L.) Garcke

Renvoize, Stoddart (1),(3)

Jef 1155(S), Fra 430(S), Rob 3125(S)

Localities recorded: S,Bir,Coe,Rem,Far

Erect hairy pantropical woody herb with ovate serrated edged leaves to 5 cm long, yellow flowers to 1 cm wide in axillary clusters, the calyx base covered in rusty hairs, and capsules to 5 mm long.

Sida L.

S. acuta Burm.f.   'la bolze, herb dur'

Summerhayes, Renvoize, Fosberg (2), Stoddart (3), Wilson (2)

Jef 624(S), Fra 451(S), Rob 2361,2362,2691,2748,3009,3167(S)

Localities recorded: M,S,F,C,Fe,Bir,Den,Pla,Rem,Poi,Pro,Ald,Ass,Cos

Woody herb with slightly zigzag stems, ovate or linear leaves to 7 cm long with short narrow stipules and scattered stellate pubescence, yellow flowers on 5 mm stalks in the leaf axils and capsules with 6 to 10 mericarps. (Fig. 13) (The *S. acuta/S. stipulata* group is difficult and may be further divided and the determinations quoted under these two species should not be relied upon.)

S. cordifolia L.

Summerhayes

Bru/War 47(S)

Localities recorded: M,P,A,L

Softly hairy woody herb with rounded to ovate leaves to 4 cm long, with long petioles, crowded yellow flowers (rarely solitary), and mericarps with awns longer than the sepals.

S. pusilla Cav. (*S. parvifolia*, *S. vescoana*)   'herb dur'

Summerhayes, Gwynne, Renvoize, Fosberg (2), Stoddart (1),(2),(3), Wilson (3),(4)

Jef 1180,1203(S), Pig sn, Pro 4228(S), Fra 439(S), Rob 2298,2354,2839,2959, 2981, 3017,3077(S)

Localities recorded: P,D,F,Bir,Den,Pla,Coe,Afr,Rem,Des,Dar,Jos,Poi,Mar,Def, Alp,Pie,Far,Ald,Ass,Cos, Ast

Spreading, often prostrate, woody herb with almost round serrated edged leaves to 2 cm long and small yellow flowers to 1.5 cm wide on long stalks singly from the leaf axils.

Fig. 13. Sida acuta

Fig. 14. Sida stipulata

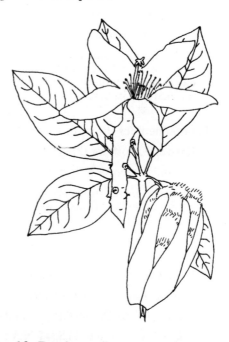

Fig. 15. Adansonia digitata

Fig. 16. Bombax ceiba

19

S. rhombifolia L.   'herb dur'

Bailey, Gwynne, Renvoize, Fosberg (2), Stoddart (1), (2), Baumer

Rob 2605,2651,2749,3197(S)

Localities recorded: M,F,N,Dar,Ald

Tall wiry herb with old stems reddish, leaves ovate to 2.5 cm long, grey beneath and on short petioles, yellow flowers on stalks to 4 cm long and capsules with blunt mericarps. Used medicinally in Seychelles.

S. stipulata Cav.

Robertson (6)

Fos 52094(S),Pro 4155(S),Rob 2420,2603,2747,2844,2958,3014,3159, 3168,3190(S)

Localities recorded: M,F,N,An,Co,Pla,Coe,Dar,Poi,Alp

Woody herb similar to *S. acuta*, but with wider leaves, denser stellate pubescence, and longer, wider stipules. (Fig. 14) (See note under *S. acuta*)

Thespesia Soland. ex Correa

T. populnea (L.)Soland. ex Correa   'bois de rose'

Summerhayes, Bailey, Gwynne, Renvoize, Sauer, Lionnet (1), Fosberg (2), Stoddart (1),(3), Baumer, Wilson (1), Robertson (4)

Ves 5658(K), Pig sn, Fos 52085(S), Fra 446(S)

Localities recorded: M,S,P,F,C,Co,Bir,Den,Rem,Pie,Ald,?Cos,?Ast

Indigenous coastal tree or shrub with heart-shaped leaves to 25 cm long, erect bright yellow flowers with a maroon centre to 9 cm long, held in a cup-shaped calyx and indehiscent fruit to 3 cm wide with seeds with silky hairs. Used medicinally in Seychelles.

T. populneoides (Roxb.)Kostel.   'bois de rose'

Renvoize, Fosberg (2)

Localities recorded: Dar,Pie,Ald,Ass,Cos,Ast

Indigenous tree or shrub with bronzy heart-shaped leaves to 15 cm long and drooping yellow flowers with red centres to 6 cm long with dehiscent fruit containing pubescent seeds.

Urena L.

U. lobata L.   'herb herisson rouge, herbe panier, aramina fibre, congo jute'

Summerhayes, Bailey, Robertson (6)

Jef 676,749(S), Pro 4025(K), Rob 2457,2595,2650(S)

Localities recorded: M,S,D,F,N,An

Cultivated and escaping woody herb with erect stems, hairy leaves to 10 cm long, sometimes deeply lobed, pinky mauve flowers to 2.5 cm wide and indehiscent 5-lobed fruit to 1.5 cm wide with short hooked spines.

BOMBACACEAE (31/2)

Adansonia L.

A. digitata L. 'baobab'

Bailey, Robertson (6)

Localities recorded: M(sr only),An,Ce(sr only)

Introduced tree with immensely wide trunk and short stout branches, leaves with 5 palmately arranged sessile leaflets, white night-opening flowers to 15 cm wide and large brown hard-shelled fruit to 20cm long containing 'cream of tartar' pith and brown seeds to 1 cm long. (Fig. 15)

Bombax L.

B. ceiba L. (*B.malabaricum*) 'ouatier rouge, silk cotton'

Bailey

No locality recorded

Large tree to 25 m with thick trunk with large pyramidal spines, leaves with 5 palmately arranged leaflets to 20 cm long, red flowers to 10cm wide and large fruit to 15 cm long containing many seeds embedded in white floss. (Fig. 16)

Ceiba Mill.

C. pentandra (L.)Gaertn. (*Eriodendron anfractuosum*) 'ouatier blanc, kapok'

Bailey, Lionnet (1),(2), Evans, Robertson (4)

Localities recorded: M,P,F

Cultivated and escaping tall soft-wooded tree, often with basal buttresses and some spines on the trunk, with horizontal branches, palmately compound leaves with 5 to 11 leaflets, white flowers, hairy outside, to 3 cm long and large pendant fruit to 15 cm long containing many seeds embedded in white floss which is used to stuff cushions, mattresses, and in the past, lifejackets.

Durio Adans.

D. zebethinus Murr. 'durian'

Bailey, Lionnet (2)

Localities recorded: M(BG)

Cultivated tall spreading tree with simple oblong leaves to 15 cm long, silvery beneath, yellowish green flowers borne on the trunk or branches, and large globose fruit to 25 cm wide with spiky exterior and strongly smelling edible pulp.

Ochroma Sw.

O. pyramidale (Cav. ex Lam.)Urb.   'balsa'

Bailey

No locality recorded

Cultivated tree to 30 m, often with basal buttresses, large angular or lobed pubescent leaves to 25 cm wide, off-white flowers to 13 cm long and fruit to 25 cm long with silky floss around the seeds. The wood is noted for its light pith-like texture.

STERCULIACEAE (32)

Cola Schott & Endl.

C. nitida (Vent.)Schott & Endl. (*C. acuminata* — Purseglove notes that *C. sp.* sent to various botanic gardens in late 1800's appear to be *C. nitida*)   'kola'

Bailey, Lionnet (2)

Ber 14663(K), Rob 2755(S)

Localities recorded: M,S

Cultivated and escaping tall tree with drooping elliptic leaves to 33 cm long, cream flowers to 3 cm long with red marks inside and lobed fruit to 13 cm long, shining green with smooth tubercles, containing several large seeds with red cotyledons. The seeds are chewed as a stimulant drug in some countries. (Fig. 17)

Guazuma Mill.

G. ulmifolia Lam. (*G.tomentosa*)   'chikrassia, bastard cedar'

Summerhayes, Swabey

Localities recorded: M

Cultivated tree with two-ranked pointed oblong serrated edged leaves to 15 cm long, unequal at the base, axillary clusters of small white, pink or yellow flowers to 5 mm long and a 5-valved woody globose fruit to 2 cm long.

Heritiera (Dryand.) Ait.

H. littoralis Ait.   'bois de table, looking glass tree'

Summerhayes, Bailey, Lionnet (1), Wilson (1),(2), Robertson (4)

Jef 503(S), Rob 3149(S)

Localities recorded: M,S,P,F,C,Coe

Indigenous coastal tree to 15 m with elliptical leaves, silvery below, to 20 cm long, inflorescences of small greenish bell-shaped flowers, each to 5 mm long, and glossy hard ovate brown fruit to 6 cm long with a prominent keel on one side.

Fig. 17. Cola nitida

Fig. 18. Theobroma cacao

Fig. 19. Waltheria indica

Fig. 20. Triumfetta rhomboidea

Melochia L.

M. betsiliensis Baker

Summerhayes

Localities recorded: M

Woody herb with triangular serrated edged leaves to 3 cm long, small yellow and violet flowers to 5 mm long and globose beaked capsules to 4 mm long.

M. pyramidata L.

Summerhayes

No locality recorded

Pantropical weed similar to *M. betsiliensis* but with ovate leaves to 4 cm long and a fat angular capsule, golden papery and shining, to 8 mm long.

Theobroma L.

T. cacao L.   'cacao, cocoa'

Bailey, Baumer, Procter (2)

Localities recorded: M(sr only),P

Cultivated tree to 8 m with large drooping oblong leaves to 60 cm long, small purplish flowers to 6 mm long borne on the trunk or branches and cylindrical red or yellow pods to 30 cm long containing rows of seeds covered in white pulp. The fermented and dried seeds yield the cocoa and chocolate of commerce. Used medicinally in Seychelles. (Fig. 18)

Waltheria L.

W. indica L. (*W.americana*)   'guimauve'

Bailey, Robertson (6)

Jef 678(S), Rob 2441,3191(S)

Localities recorded: M,An

Pantropical weed, an erect pubescent herb to 1 m with prominently nerved ovate leaves to 9 cm long and dense axillary heads of small yellow flowers, each to 5 mm long. (Fig. 19)

TILIACEAE (33/1)

Berrya Roxb.

B. amonilla Roxb.   'berrya, trincomali wood'

Bailey

No locality recorded

Timber tree with heart-shaped leaves to 25 cm long or more, white flowers to 5 mm long in spreading inflorescences and golden brown capsules with 6 oblong wings, each to 3 cm long.

Corchorus L.

C. aestuans L. (*C. acutangulus*)

Summerhayes, Renvoize, Fosberg (2)

Localities recorded: M,?Dar,Ald,Ass,Cos,Ast

Spreading woody hairy herb with pointed ovate serrated edged leaves to 5 cm long, yellow flowers to 4 mm long and longitudinally ridged capsule to 2.5 cm long with 3 to 5 beaks at the apex and many small brown seeds.

Grewia L.

G. aldabrensis Baker

Renvoize, Fosberg (2)

Localities recorded: Ald

Endemic slender tree or shrub with ovate serrated edged leaves to 7cm long, with 3 main veins, yellow flowers to 6mm long and small glabrous 4-lobed fruits to 5mm long.

G. asiatica L.   'napoleon blanc'

Rob 3204(S)

Localities recorded: M

Tree or shrub with reddish net veined branches, rounded, serrated edged leaves to 18cm long, pointed at the apex and with 5 main veins from the base, yellow flowers and black fleshy globose fruit to 1cm long.

G. salicifolia Schinz   'mabolo'

Renvoize, Fosberg (2)

Localities recorded: Ald,Cos,Ast

Endemic small to medium tree with lanceolate serrated edged leaves to 18cm long, purplish pink flowers to 1cm long, the sepals longer than the petals, and hairy woody 4-lobed fruits to 2cm wide.

Triumfetta L.

T. procumbens Forst.f. (*T. repens*)

Summerhayes, Gwynne, Renvoize, Fosberg (2), Stoddart (1),(2)

Rob 2388,2864,3056(S)

Localities recorded: M,Coe,Des,Dar,Poi,Alp,Pro,Far,Ast

Indigenous densely hairy prostrate strand plant with simple or 3 lobed leaves to 5 cm long, yellow flowers to 1 cm long and globose fruit to 1.5 cm wide covered in hooked spines.

T. rhomboidea Jacq. (*T. bartramia*)   'herb herisson blanc, herb panier'

Summerhayes, Bailey, Baumer

Jef 677(S), Rob 2585,2677,2833(S)

Localities recorded: M,S,F,N,L,Alp

Pantropical weed, a bushy herb with ovate or rhomboid leaves to 8 cm long, axillary clusters of small yellow flowers and globose fruit to 5 mm wide covered in hooked spines. Used medicinally in Seychelles. (Fig. 20)

ERYTHROXYLACEAE (34/2)

Erythroxylum P.Br.

E. acranthum Hemsley   'sandol, bois sandol'

Renvoize, Fosberg (2)

Localities recorded: Ald,Ass,Cos

Endemic shrub to 5 m with obovate leathery leaves to 5 cm long, axillary clusters of cream flowers, each to 4 mm wide and red oblong fruit to 4mm long.

E. longifolium Lam. vel. aff.

Ves 5521,5546(K), Rob 2518,2767,3173,3196(S)

Localities recorded: S,An,T,Ce

Glossy leaved shrub similar to *E.sechellarum* but with pointed leaves, horizontal branches and flowers and fruit on short stalks.

E. sechellarum O.E.Schultz   'bois de ronde, cafe marron petite feuille'

Summerhayes, Bailey, Procter (1)

Ves 5423,5432(K), Jef 462,472,481,482,644(S),668(K), Sch 11660,11827(K), Fos 51996(S), Pro 4523,4537(K), Ber 14620(K), Rob 2276,2406,2704,2804(S)

Localities recorded: M,S,P,D,C,Fe

Endemic glossy leaved shrub with leathery leaves to 12 cm long, solitary white axillary flowers to 5 mm long and shiny oblong red fruit to 1 cm long with one seed.

MALPIGHIACEAE (36)

Galphimia Cav.

G. gracilis Benth. (Jeffrey's *Thryallis glauca*)

Jef sn(K)

Localities recorded: M

Cultivated shrub to 2 m with ovate, opposite leaves to 6 cm long and loose terminal sprays of yellow flowers to 1.5 cm wide, the petals with a narrow base.

Malpighia L.

M. glabra L.   'barbados cherry'

Bailey

Localities recorded: M(sr only)

Cultivated shrub to 4 m with dark green ovate opposite leaves to 10 cm long, pinkish red frilly petalled flowers to 1.5 cm wide and red globose edible fruit to 2 cm wide containing 3 seeds in flesh rich in Vit C.

Tristellateia Thouars

T. australasiae A.Rich.

Localities recorded: M(sr only)

Cultivated climber with ovate opposite leaves to 14 cm long, loose terminal sprays of yellow flowers to 2.5 cm wide and flat star-shaped brown fruit to 1 cm wide.

ZYGOPHYLLACEAE (37)

Tribulus L.

T. cistoides L. 'pagode'

Summerhayes (not *T.terrestris*), Gwynne, Renvoize, Fosberg (2), Stoddart (1),(3), Wilson (3)

Fra 427 (S)

Localities recorded: Den,Bir,Afr,Rem,Def,Far,Ald, Ass,Cos,Ast

Pantropical weed, a prostrate herb with opposite pinnate leaves to 7 cm long, yellow flowers to 2 cm wide and fruit with 5 sections each bearing 4 stout spines.

GERANIACEAE (38/1)

Pelargonium L'Hér.

P. zonale (L.)L'Hér. ex Ait. 'geranium'

Bailey

Localities recorded: M(sr only)

Cultivated woody herb with round pubescent leaves to 10 cm wide and heads of showy flowers, usually red, each flower to 3 cm wide.

TROPAEOLIACEAE (38/2)

Tropaeolium L.

T. majus L. 'nasturtium'

Bailey

No locality recorded

Cultivated climbing annual herb with round peltate leaves to 15 cm wide, yellow, orange or red flowers to 8 cm wide with one petal forming a spur and wrinkled green 3-lobed fruit to 1 cm wide.

OXALIDACEAE (38/6)

Oxalis L.

O. corniculata L.

Summerhayes (including Horne's doubtful *O.radicosa*), ?Wilson (1)

Rob 2717(S)

Localities recorded: M,S,?P,F

Pantropical weed, a herb with creeping stems, trifoliolate leaves, each leaflet heart-shaped to 1 cm long, yellow flowers and narrow capsules to 1.5 cm long which dehisce to scatter the many small brown seeds. (Fig. 21)

O. sp. near **bakerana** Exell

Fosberg (2)

Localities recorded: Cos

No description

AVERRHOACEAE (38/7)

Averrhoa L.

A. bilimbi L.   'bilimbi, cucumber tree'

Bailey, Lionnet (1), Baumer, Wilson (1),(2), Robertson (2),(4)

Fos 52112(S), Bru/War 4(S), Rob 2462,2583(S)

Localities recorded: M,P,F,N,C,A,Co,Coe

Cultivated and escaping small tree with softly hairy pinnate leaves with up to 36 oblong leaflets each to 12 cm long, clusters of small red flowers, each to 2 cm long, borne on the trunk and branches, and yellowish green acidic edible cucumber shaped fruit to 10 cm long. Used medicinally in Seychelles.

A. carambola L.   'carambolier'

Bailey, Lionnet (1),(2)

Jef 608(S), Sch 11703(K), Pro 4467(K)

Localities recorded: M

Cultivated and escaping small densely branched tree to 7 m with glabrous pinnate leaves with up to 20 ovate leaflets each to 8 cm long, axillary clusters of red and white flowers and yellow fleshy edible oblong fruit to 10 cm long, markedly 5 angled. (Fig. 22)

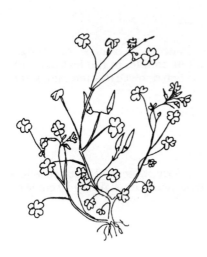

Fig. 21. Oxalis corniculata

Fig. 22. Averrhoa carambola

Fig. 23. Impatiens walleriana

Fig. 24. Murraya paniculata

BALSAMINACEAE (38/8)

Impatiens L.

I. balsamina L.   'balsam'

Summerhayes

Localities recorded: M

Cultivated annual herb, erect and fleshy stemmed, with ovate serrated edged leaves to 15 cm long, colourful flowers, often double, in the leaf axils and pointed ovoid green capsules to 2 cm long which dehisce to scatter the seeds.

I. gordonii Horne ex Baker   'balsamine sauvage'

Summerhayes, Bailey, Procter(1)

Localities recorded: M,?S (probably extinct)

Endemic fleshy herb with long-stalked narrowly ovate and serrated edged leaves to 15 cm long and flowers with narrow petals on stalks to 10 cm long.

I. thomassetii Hook.f.   'balsamine sauvage'

Summerhayes, Bailey, Procter(1)

Pro 3972(S)

Localities recorded: M

Endemic fleshy herb with ovate serrated edged leaves to 20 cm long and white flowers to 6 cm wide with a pink spur.

I. walleriana Hook.f.   'balsamine, busy lizzie'

Bailey

Jef 435,491(S)

Localities recorded: M

Cultivated fleshy plant with ovate tooth-edged leaves to 12 cm long and showy single pink or white flowers to 4 cm wide. (Fig.23)

RUTACEAE (39)

Aegle Correa

A. marmelos Correa   'bael fruit'

Bailey

No recorded locality

Cultivated thorny tree to 3 m with trifoliolate gland-dotted leaves, the leaflets crenate and to 7 cm long, fragrant white flowers and greyish yellow smooth hard globose fruits to 17.5 cm wide containing edible flesh which becomes hard and red on drying.

Citrus L.

(The following names have been taken from Purseglove — there is much confusion)

C. aurantifolia (Christm.)Swing.   'limon (Cr), lime (Eng)'

Fosberg (1),(2), Stoddart (3), Robertson (2),(3),(4), (5),(7)

Bru/War 91(S)

Localities recorded: M(sr only),F,N,A,Co,Bir,Den,Dar,Poi,Coe,Ald

Cultivated much branched small spiny tree to 5 m with shiny gland-dotted leaves to 8 cm long, the petioles narrowly winged, white flowers and fleshy globose edible fruit to 6 cm wide, skin yellow when ripe and the flesh greenish and acid.

C. aurantium L. (*C. vulgaris*)   'oranger, ?orange mozambique (Cr), seville orange(Eng)'

Bailey, Robertson (4),(5)

Localities recorded: M(sr only),F,N

Cultivated medium-sized tree to 10 m with slender spines, shiny gland-dotted leaves to 10 cm long, the petioles broadly winged, large white flowers and fleshy subglobose fruit to 10 cm wide, with skin bright orange red when ripe and bitter flesh, the central core hollow.

C. aurantium L. ssp. **bergamia** (Risso & Poit.) Wight & Arn. (*C.bergamier*) 'bergamotier'

Bailey

No locality recorded

Cultivated tree similar to *C.aurantium* but with smaller flowers and fruit, the fruit globose to pear-shaped and bright yellow.

C. aurantium L. ssp. **bigaradier** L. var. **mitis** (not a Purseglove name) 'bigaradier'

Bailey, Fosberg (1), Baumer, Robertson (5)

Localities recorded: M(sr only),N,Co

Cultivated tree similar to *C.aurantium* but with smaller leaves and the petioles not winged, and the fruit similar to *C.aurantifolia* but bright orange. Used medicinally in Seychelles.

C. grandis (L.)Osbeck (*C.decumana*)   'pummelo, ?pamplemousse (Cr), shadock (Eng)'

Bailey

No locality recorded

Cultivated spreading spiny tree to 15 m with shiny gland-dotted ovate leaves to 20 cm long, the leaf petioles broadly winged, large cream flowers and large globose fruits to 30 cm wide, the thick skin yellowish when ripe with pale yellow or pink edible flesh.

C. hystrix DC. 'combava'

Bailey, Robertson (4)

Localities recorded: M(sr only, research station),F

Cultivated tree used as rootstock for other *Citrus* species, the petiole-wings almost as wide as the leaf-blade, the fruit ?edible.

C. limon (L.)Burm.f. (*C. acida*)   'limonier, gros limon, ?citron (Cr), lemon (Eng)'

Bailey, Baumer, Robertson (4),(5)

Bru/War 94(S)

Localities recorded: M(sr only),F,N,A

Cultivated tree to 6 m with stout spines, shiny gland dotted leaves to 10 cm long, the petioles not winged, flowers pinkish purple in bud and opening white, and ovoid fruit to 10 cm long with a terminal nipple, the skin yellow when ripe and the edible flesh pale yellow and sour. Used medicinally in Seychelles.

C. medica L.   'citronnier (Cr), citron (Eng)'

Bailey

No locality recorded

Cultivated tree to 3 m with stout spines, shiny gland dotted leaves to 20 cm long, the petioles short and not winged, pink tinged flowers and large oblong fruit to 20 cm long with bumpy thick yellow skin and sour greenish edible pulp.

C. paradisi Macf. (*C. decumana* var. *paradisi*)   '?pamplemousse (Cr), grapefruit (Eng)'

Bailey, Robertson (3),(4),(5)

Localities recorded: M(sr only),F,N,Poi

Cultivated spreading tree to 15 m with shiny gland-dotted leaves to 15 cm long, the petioles broadly winged, white flowers and large flattened globose fruit to 15cm wide with smooth pale yellow skin and edible juicy flesh, pale greenish yellow or pink.

C. reticulata Blanco (*C. nobilis, C. vangasay*)   'mandarinier, vengasaille (Cr), tangerine, mandarin, satsooma (Eng)'

Bailey, Robertson (4),(5)

Bru/War 92(S)

Localities recorded: M(sr only),F,N,A

Cultivated, sometimes spiny, tree to 8 m with shiny gland-dotted leaves to 8 cm long, the petioles margined but not winged, white flowers and depressed globose fruit to 8 cm wide, the skin greeny yellow to bright orange red and easily detachable and the edible flesh juicy sweet and orange.

C. sinensis (L.)Osbeck 'orange'

Fosberg (1), Robertson (3),(4),(5)

Bru/War 90,95(S)

Localities recorded: M(sr only),F,N,A,Co

Cultivated tree to 12 m with angled twigs and sometimes with spines, gland-dotted shiny leaves to 15 cm long, the petioles narrowly winged, white flowers and globose fruit to 12 cm wide, the skin tightly adhering and pitted, ripening orange in temperate regions, but remaining green in the tropics, with juicy sweet edible orange flesh.

Clausena Burm.f.

C. anisata (Willd.)Oliv. 'carri pile'

Bailey

Jef 569,570(S)

Localities recorded: M

Cultivated small glabrous tree with pinnate gland-dotted leaves, the leaflets one-sided at the base and to 10 cm long, greenish white fragrant flowers in clusters and globose or ovoid fruit to 1 cm long, whitish when ripe. In Seychelles the leaves are used as a pot herb.

Murraya Koen. ex L.

M. paniculata L. 'bois buis'

Jef 1157(K), Fos 52200(S), Pro 4005,4027,4378(S), Rob 2766(S)

Localities recorded: S,D,Dar

Cultivated and escaping shrub or small tree with dark green gland-dotted pinnate leaves, the leaflets obovate and to 4 cm long, fragrant white flowers to 1.5 cm long and orange-red ovate fruit to 1 cm long. (Fig. 24)

Toddalia Juss.

T.sp. 'patte de poule'

Bailey

No locality recorded

No description (*T. asiatica* is a scrambling prickly shrub with trifoliolate leaves, creamy green flowers to 3mm long and yellow pea-sized hot tasting fleshy fruit).

Triphasia Lour.

T. trifolia (Burm.f.)P.Wilson   'orangine'

Rob 3192(S)

Localities recorded: M

Cultivated spiny shrub with greyish trifoliolate leaves, the leaflets sinuate edged and the largest to 4 cm long, small white flowers and orange-red ovoid fruit to 8 mm long.

SIMAROUBACEAE (40/1)

Brucea J.F.Mill.

B. javanica (L.)Merr.

Jef 363(S), Pro 4214(S), Rob 2541(S)

Localities recorded: M

Shrub with odd pinnate leaves to 40 cm long, the leaflets serrated edges and pointed ovate, spikes of small yellowish green flowers with red anthers in the leaf axils and small ovoid black fruit to 7 mm long.

Quassia L.

Q. amara L.   'quassia, bitter wood'

Bailey

Rob 3206(S)

Localities recorded: M

Erect glabrous shrub to 5 m with pinnate leaves to 25 cm long with winged rachis, terminal spikes of narrow red flowers, each to 5 cm long with the stamens exserted, and small ovoid fruits. A medicinal plant in some countries. (Fig. 25)

Soulamea Lam.

S. terminalioides Baker   'colophante'

Summerhayes, Bailey, Procter (1)

Ves 5534,6107(K),Jef 572,645(S),Sch 11830(K),Fos 51968(S),Pro 3939,4239(S), Ber 14694(K), Rob 2529(S)

Localities recorded: M,S(formerly)

Endemic tree with oblong leaves to 15 cm long with long petioles, clustered at the ends of branches, spikes of small yellow flowers, ♂ and ♀ on separate trees, and three-winged fruits to 2.5cm long. (Fig. 26)

SURIANACEAE (40/3)

Suriana L.

S. maritima L.   'bois d'amande, matelot, bois chevre'

Fig. 25. Quassia amara

Fig. 26. Soulamea terminalioides

Fig. 27. Suriana maritima

Fig. 28. Melia azederach

Summerhayes, Bailey, Gwynne, Renvoize, Sauer, Fosberg (2), Stoddart (1),(2),(3), Baumer

Fos 52065(S), Pro 4457,4496(K), War 183(S), Rob 2358, 2846,3030,3067(S)

Localities recorded: P,C,A,Ce,Co,Bir,Den,Pla,Coe,Afr,Rem,Des,Dar,Jos, Poi,Alp,Pro,Far,Ald,Ass,Cos,Ast

Indigenous much branched strand shrub with greyish tomentose branches and narrow obovate leaves to 3cm in terminal clusters, small yellow flowers and fruit of 5 globose sections, each to 3mm wide, held in the calyx. Used medicinally in Seychelles. (Fig. 27)

OCHNACEAE (41/1)

Ochna L.

O. ciliata Lam.   'bois mangu, bois bouquet'

Bailey(probably his *O.thomasiana)*, Renvoize, Fosberg (2)

Pro 4279,4532(K)(?*O.kirkii)*

Localities recorded: M,Ald

Shrub or tree with shiny dark green ovate leaves to 12 cm long with tiny spines on the lower edges, fragrant yellow flowers with the petals falling early and the sepals turning red and reflexed to show the black ovoid 1cm long fruits attached to the red receptacle.

BURSERACEAE (42)

Canarium L.

C. commune L.   'amande java'

Bailey

No locality recorded

Tree with odd-pinnate leaves, the leaflets pointed ovate to 15 cm long, and crowded small flowers. The source of the resin 'manila elemi'.

MELIACEAE (43)

Cedrela P.Br.

C. odorata L.   'west indian cedar'

Swabey

Localities recorded: M

Cultivated timber tree to 30 m with odd-pinnate leaves to 70cm long, the ovate leaflets to 15 cm long, lax sprays of small yellowish flowers and 5 valved oblong capsules to 3 cm long with seeds with a wing at one end.

Khaya A.Juss.

K. senegalensis A.Juss.   'mahogany'

Bailey, Swabey (?*K.nyassica*)

Localities recorded: M

Cultivated timber tree to 30 m with odd-pinnate leaves to 25 cm long, the leaflets oblong-elliptic to 10 cm long, loose sprays of white flowers and 4-valved woody capsules with winged seeds.

Lansium Correa

L. domesticum Jack   'langsat'

Bailey

No locality recorded

Tree with odd-pinnate leaves with alternate ovate leaflets to 15 cm long, small sessile flowers in greyish spikes on trunk and branches and oblong fruit to 4 cm long with tough greyish skin with latex canals and edible translucent pulp with one or two seeds.

Malleastrum (Baill.)Leroy

M. leroyi Fosberg   'bois trois feuilles'

Renvoize, Fosberg (2)

Localities recorded: Ald

Small endemic tree with trifoliolate leaves, the leaflets ovate and to 11 cm long, loose heads of small fragrant flowers, ♂ and ♀ on separate plants, and ovoid fruit.

Melia L.

M. azederach L.   'lilas d'inde, persian lilac'

Summerhayes, Bailey

Localities recorded: M

Tree to 10 m with bipinnate leaves to 1 m long, large loose sprays of pale lilac fragrant flowers and globose yellow fruit to 1 cm wide. (Fig. 28)

M. dubia Cav.   'lilas'

Bailey, Swabey, Wilson (1)

Rob 2690(S)

Localities recorded: S,P,F

Cultivated and escaping tree to 15 m with bipinnate leaves to 1 m long, the leaflets ovate and tooth edged to 7 cm long, sprays of many small pinkish white flowers, each to 1 cm long, and fleshy ovoid yellow fruit to 3 cm long.

Sandoricum Cav.

S. koetjape (Burm.f.)Merr. (*S.indicum*)   'santol'

Bailey, Lionnet (1), Wilson (1), Robertson (5)

Rob 2237(S)

Localities recorded: M,P,N

Cultivated and escaping tree with pubescent young branches and leaves, the leaves trifoliolate with wide ovate leaflets to 20 cm long, sprays of yellowish flowers and globose pubescent pale orange fruit to 7 cm wide containing a few large seeds surrounded by edible pulp.

S. radiatum King   'santol'

Bailey

No locality recorded

Similar to *S. koetjape*. It is doubtful whether Bailey did identify this species as present.

Swietenia Jacq.

S. macrophylla King   'mahogany, honduras mahogany'

Bailey, Swabey, Wilson (1), Robertson (4)

Localities recorded: M,P,S,F

Cultivated timber tree to 50 m with pinnate leaves to 30 cm long, the uneven-based leaflets to 20 cm long, axillary sprays of small yellowish flowers, each to 5 mm long, and oblong woody 5-celled fruit to 15 cm long containing winged seeds to 7 cm long. Widely planted as a valuable timber crop on Mahe.

S. mahagoni Jacq.   'mahogany, west indian mahogany'

Bailey, Wilson (1)

Localities recorded: P

Cultivated timber tree to 25 m with pinnate leaves to 18 cm long, the leaflets to 8 cm long, axillary sprays of small flowers, each to 5 mm long, and woody oblong fruit to 12 cm long containing winged seeds to 5 cm long.

Xylocarpus Koen.

X. granatum Koen. (*Carapa obovata*)   'manglier, manglier pomme, ?casse tete chinois, puzzle nut'

Summerhayes, Bailey, Lionnet (1), Renvoize, Fosberg (2), Wilson (2), Robertson (4)

Localities recorded: M,P,F,C,Ald,Cos

Indigenous spreading small tree or mangrove with pinnate leaves with 3 to 5 obovate rounded leaflets to 9 cm long, fragrant creamy yellow flowers to 5 mm long held at an angle to the stalk and large glossy orange brown globose fruit to 18 cm wide containing a number of irregularly angled seeds. (Fig. 29)

Fig. 29. Xylocarpus granatum   Fig. 30. Ximenia americana

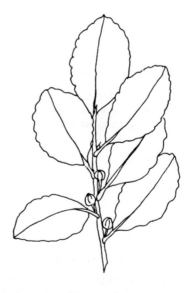

Fig. 31. Apodytes dimidiata   Fig. 32. Ilex paraguariensis

X. moluccensis (Lam.)Roem.   'manglier pomme'

Summerhayes, Renvoize, Sauer, Fosberg (2)

Pro 4261(K)

Localities recorded: M,P,C,Ald,Cos

Indigenous small tree or mangrove with pinnate leaves with 3 to 7 pointed ovate leaflets, each to 10 cm long, white flowers at an angle to their stalks and globose glossy orange brown fruit to 15 cm wide splitting into 4 valves containing large irregular seeds.

OLACACEAE (45/1)

Ximenia L.

X. americana L.

Summerhayes

Localities recorded: P

Indigenous small tree to 7 m with axillary spines, oblong leaves to 8 cm long, greenish white fragrant flowers to 1 cm long and hairy inside and ovoid yellow fleshy fruit to 2.5 cm long with one hard seed. (Fig. 30)

ICACINACEAE (45/3)

Apodytes E.Mey. ex Arn.

A. dimidiata E.Mey. ex Arn.   'bois marie, la fouche petit feuille, bois none'

Renvoize, Fosberg (2)

Localities recorded: M(BG),Ald

Small tree to 5 m with ovate leathery leaves to 10 cm long with red petioles, sprays of small white scented flowers and asymmetrical small flattened black fruit to 1cm long with a red lobe. (Fig. 31)

Grisollea Baill.

G. thomassetii Hemsley

Summerhayes, Bailey, Procter (1)

Dup 11/32(K), Ves 5548,6192,6201,6330(K), Jef 697,807 (S), Pro 4012,4015,4071(S), Ber 14642,14643(K)

Localities recorded: M,S

Endemic medium sized tree with oblong leaves to 25 cm long, flowers very small in stalked heads, ♂ and ♀ on separate trees, and pointed and flattened ovate fruit to 2 cm long.

AQUIFOLIACEAE (46/1)

Ilex L.

I. paraguariensis A.St-Hil.   'thé du mexique, yerba de maté'

Bailey

No locality recorded

Tree with obovate alternate crenately edged leaves to 10 cm long, inconspicuous axillary greenish white flowers and small 4-ribbed black fruits to 6 mm long. Used as a herbal plant in some countries. (Fig. 32)

CELASTRACEAE (47/1)

Elaeodendron Jacq.

E. orientale (Jacq.)O.Kuntze (*Cassine orientalis*)   'bois d'olive'

Bailey

Jef 588(S), Dup 5/18(K)

Localities recorded: M(BG)

Cultivated small tree with ovate leaves to 10 cm long, axillary clusters of small greenish flowers, each to 5 mm wide, and ovoid orange fruit to 1.5 cm long.

Maytenus Molina

M. senegalensis (Lam.)Exell

Renvoize, Fosberg (2)

Localities recorded: Ald,Ass,Cos,Ast

Small, sometimes spiny tree with pale green leathery obovate leaves to 5 cm long, clusters of small cream flowers and reddish brown dehiscent capsules to 4 mm wide containing 4 glossy red-brown seeds with fleshy rose-pink arils. (Fig. 33).

Mystroxylon Eckl. & Zeyh.

M. aethiopicum (Thunb.)Loes.   'ti bane, bois moset'

Renvoize, Fosberg (2)

Localities recorded: Ald,Ass,Ast

Small bushy tree with ovate leaves to 7 cm long, axillary clusters of minute yellow flowers and small red ovoid single seeded edible fruit to 1 cm long.

RHAMNACEAE (49)

Colubrina Brongn.

C. asiatica (L.)Brongn.   'bois savon, savonnier'

Summerhayes, Bailey, Renvoize, Fosberg (2), Wilson (2),(3), Robertson (4),(6)

Per/Rid 1955(K), Jef 834(S), Pro 4014,4231(S), Rob 2408,2789,3004(S)

Localities recorded: M,S,P,F,C,An,Pla,Def,Ald,Cos,Ast

Indigenous shrub with long lax branches, ovate pointed serrated edged leaves to 8 cm long, axillary clusters of small greenish flowers and faintly 3-lobed fruit to 8 mm wide, green when young and brown when ripe, on long stalks. (Fig. 34)

Gouania Jacq.

G. scandens (Gaertn.)R.B.Drummond   'liane charretiers'

Renvoize, Fosberg (2)

Localities recorded: Ald,Cos

Scrambling twining shrub with reddish hairs on young stems, pointed broadly ovate leaves to 7 cm long and a few minute flowers on spikes which elongate to bear fruit to 1 cm long with 3 wings.

Scutia Comm. ex Brongn.

S. myrtina (Burm.f.)Kurz   'tire bonnet, bambara, bois senti'

Renvoize, Fosberg (2)

Localities recorded: Ald,Ass,Cos,Ast

Spreading shrub with small recurved thorns, rounded or obovate leaves to 6 cm long, inconspicuous axillary flowers subtended by tendrils and globose black fleshy fruit to 8 mm wide with 2 seeds, held in a cup-like calyx. (Fig. 35)

Smythea Seem.

S. lanceata (Tul.)Summerhayes

Summerhayes

Localities recorded: M.P

Indigenous climbing shrub, with zigzag stems, pointed oblong leaves to 15 cm long, small axillary greenish yellow flowers and woody flattened pointed ovate fruit to 5 cm long.

Ziziphus Tourn. ex L.

Z. mauritiana Lam. (*Z.jujuba*)   'jujubier, masson'

Summerhayes, Bailey (including his *Z.vulgaris*)

Jef 745(S), Rob 3195(S)

Localities recorded: M

Small spreading spiny tree with rounded leaves to 7 cm long, pale buff pubescent below with three main veins, greenish axillary flowers to 5 mm wide and globose yellow edible one-seeded fruit to 2.5 cm wide. (Fig. 36)

Fig. 33. Maytenus senegalensis   Fig. 34. Colubrina asiatica

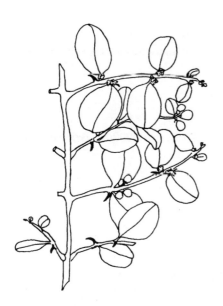

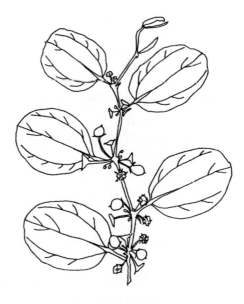

Fig. 35. Scutia myrtina   Fig. 36. Ziziphus mauritiana

VITACEAE (50/1)

Leea L.

L.sp.

Rob 2271(S)

Localities recorded: M

Shrub or creeper with compound leaves with reddish stem and flat clusters of red fruit.

Vitis L.

V. vinifera L.   'raisin, vigne, grapevine'

Bailey

Localities recorded: M(sr only)

Cultivated climber with tendrils, 3–5-lobed leaves to 20 cm wide, axillary clusters of small greenish yellow flowers and edible globose fleshy purple or golden fruit to 2 cm wide.

SAPINDACEAE (51/1)

Allophylus L.

A. aldabricus Radlk.

Renvoize, Fosberg (2)

Localities recorded: Ald,Ass,Cos,Ast

Endemic shrub to 4m with trifoliolate leaves, the largest leaflet obovate to 7cm long, spikes of minute sessile flowers and ovoid glabrous orange-red fruit to 6 mm long.

A. pervillei Blume   '?bois cafoul, bois maris'

Summerhayes

Ves 5774,6187(K), Jef 615,1152(S), Fos 52197(S), Pro 4000(S), Ber 14649(K), Rob 2811(S)

Localities recorded: M,S,P,D,Fe

Indigenous small tree with elliptic, sinuate edged leaves to 20 cm long, small greenish flowers in spikes and globose fruit to 7 mm long.

A. sechellensis Summerhayes (?*A. gardineri*)

Summerhayes, Procter (1)

Ves 6191(K), Pro 3999(S), 4368,4381,4482,4469(K)

Localities recorded: M,S

Endemic shrub to 2 m with trifoliolate leaves, the faintly toothed ovate leaflets to 15 cm long, and spikes of small white flowers which are sometimes reflexed.

Cardiospermum L.

C. halicacabum L.   'balloon vine'

Summerhayes, Stoddart (3)

Jef 838(S), Rob 2468,2596(S)

Localities recorded: M,S,F,N,Bir

Slender, sometimes woody, vine with toothed biternate leaves to 10 cm or more long, small white long stalked flowers with tendrils below and large bladder-like 3-angled capsules to 2 cm wide containing 3 black seeds with white arils.

Dodonaea Mill.

D. viscosa Jacq.   'bois de reinette'

Summerhayes, Bailey, Sauer, Fosberg (2)

Jef 778(K),1225(S), Pro 3948(S),4500(K), Rob 3116(S)

Localities recorded: M,S,P,D,C,L,Coe,Ald

Indigenous shrub with erect bright green sticky lanceolate leaves to 15 cm long, small greenish flowers in terminal clusters and pale brown papery winged fruit to 2 cm wide, ripening reddish. (Fig. 37)

Litchi Sonn.

L. chinensis Sonn. (*Nephelium litchi*)   'litchi'

Bailey

Localities recorded: M(sr only)

Cultivated tree to 10 m with glossy pinnate leaves, with 2 to 4 pairs of leaflets each to 15 cm long, sprays of pale green flowers and red tubercular globose fruit to 3 cm wide with one brown seed embedded in a white fleshy edible aril.

Macphersonia Blume

M. hildebrantii O.Hoffm.   '?tamarind marron, ?tamarin batard'

Renvoize, Fosberg (2)

Localities recorded: Ald,Ass,Cos,Ast

Tree to 4 m with bipinnate leaves to 25 cm long with oblong leaflets, axillary racemes to 10cm long of small flowers and dark purple or black hanging fruit to 1 cm long.

Nephelium L.

N. lappaceum L.   'litchi plume, rambuttan'

Bailey, Lionnet (2)

Localities recorded: M(BG)

Cultivated spreading tree to 20 m with pinnate leaves with 2 to 4 pairs of leaflets to 20 cm long, axillary clusters of small flowers, ♂ and ♀ on separate trees, and hanging red or yellow oblong fruit to 6 cm long covered in long soft spines and containing one black seed covered in a white edible aril.

Sapindus L.

S. rarak DC.

Sumerhayes

Localities recorded: M

Tree with pinnate leaves, the lanceolate leaflets to 12 cm long, branched sprays of yellow flowers and unequally lobed ovoid fruit to 8 mm wide.

S. saponaria L.   'soap berry tree'

Bailey

Nichols (K)

Localities recorded: M(BG)

Tall tree with pinnate leaves, the ovate to lanceolate leaflets to 10 cm long, small greenish white flowers in large terminal sprays and bunches of brown hollow globose fruit to 1.5 cm wide, with one black seed. Can be used as a soap substitute.

ANACARDIACEAE (53/1)

Anacardium L.

A. occidentale L.   'acajou, cashew'

Summerhayes, Bailey, Lionnet (1), Sauer, Robertson(5)

Jef 625(s), Rob 2485,2666(S)

Localitites recorded: M,P,F,N,C,Fe

Cultivated and escaping spreading tree to 12 m with leathery obovate leaves to 20 cm long, loose terminal sprays of pink flowers, each to 1 cm long, ♂ and ♀ in the same spray, and kidney-shaped nut to 3 cm long hanging beneath the swollen red or yellow receptacle or 'apple'. The skin of the nut contains a corrosive acid which has to be driven off before the nut can be opened to release the edible kernel. (Fig. 38)

Campnosperma Thw.

C. seychellarum March   'bois de montagne, ?capucin blanc, ?colophante'

Summerhayes, Bailey, Procter (1)

Ves 5482(K), Jef 542,766,1215(S), Fos 51995(S), Pro 4560(K), Rob 2782(S)

Localities recorded: M

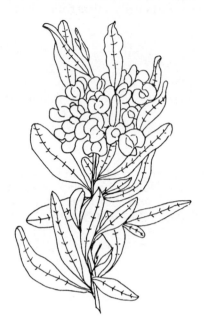

Fig. 37. Dodonaea viscosa

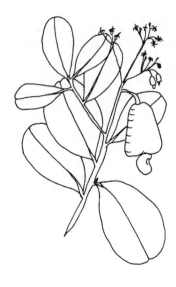

Fig. 38. Anacardium occidentale

Fig. 39. Moringa oleifera

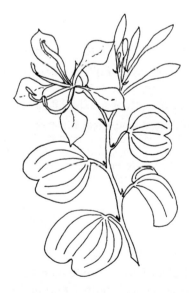

Fig. 40. Bauhinia galpinii

Endemic tree with spreading crown, lanceolate leathery leaves to 30 cm long, short terminal spikes of minute white flowers and small conical fruit to 8 mm long.

Mangifera L.

M. indica L. 'manguier, mango'

Summerhayes, Bailey, Lionnet (1),(2), Procter (2), Wilson (2), Robertson (4),(5),(6)

Fos 52113(S)

Localities recorded: M,S,P,F,N,C,An,Co

Cultivated and escaping large densely crowned tree to 40 m with lanceolate leathery leaves to 40 cm long, red when young, crowded terminal inflorescences of small brown yellow flowers, ♂ and ♀ together, and large ovoid hanging green or yellow fruit to 30 cm long with juicy edible yellow flesh around the one large flat stone.

Operculicarya Perr.

O. gummifera (Sprague)Capuron

Renvoize, Fosberg (2)

Localities recorded: Ald

Small tree to 5 m with red branches, odd-pinnate leaves with ovate leaflets to 10 cm long, spikes of small flowers coming before the leaves, ♂ and ♀ on separate trees, and small purplish ovoid fruit to 8 mm long.

Spondias L.

S. cytherea Sonn. (*S.dulcis*) 'fruit de cythere, golden apple'

Summerhayes, Bailey, Fosberg (1), Lionnet (1), Wilson (1), (2), Robertson (3),(4),(5)

Localities recorded: M,S,P,F,N,C,Co,Poi

Cultivated and escaping tree to 20 m with odd-pinnate leaves, the elliptic leaflets to 7 cm long, crowded at the ends of the branches and turning yellow before falling, small white flowers in dense terminal clusters before the leaves and hanging egg-shaped brownish yellow fruit to 8 cm long with the edible yellow flesh mixed with the strong woody fibres of the stone.

MORINGACEAE (55)

Moringa Adans.

M. oleifera Lam. 'brede morongue, horse radish tree'

Summerhayes, Bailey, Fosberg (2), Stoddart (1),(2), (3), Baumer, Wilson (4), Robertson (2),(3),(4),(6)

Sch 11842(K), Fos 52131(S), Bru/War 49(S), Rob 2888 (S)

Localities recorded: M,F,A,An,L,Co,Bir,Den,Coe,Rem, Des,Dar,Jos,Poi,Mar, Alp,Far,Ald,Ass,Cos,Ast

Cultivated and escaping tree with compoundly pinnate leaves with many ovate leaflets each to 1 cm long, sprays of white pea-like flowers each to 1.5 cm long and cylindrical ridged narrow pods to 40 cm long splitting longitudinally when ripe into 3 to release the brown winged seeds. The leaves are used as spinach and in some countries the young pods are used as a vegetable. Used medicinally in Seychelles. (Fig. 39)

LEGUMINOSAE: CAESALPINIOIDEAE (57)

Amherstia Wall.

A. nobilis Wall.   'amherstia'

Bailey

No locality recorded

Ornamental tree with pinnate leaves, the ovate pointed leaflets to 20 cm long, drooping bunches of young leaves pink, hanging axillary sprays of striking pale red flowers to 10 cm wide with persistent red bracts to 7 cm long and flat brown pods.

Bauhinia L.

B. galpinii N.E.Br.   'bauhinia rouge, sabot boeuf rouge, bois de boeuf rouge'

Bailey

No locality recorded

Small ornamental tree with wide bilobed leaves to 7 cm long, brick red flowers to 8 cm wide and hard brown flat pods to 12 cm long containing dark brown seeds. (Fig. 40)

B. variegata L.   'bauhinia, sabot boeuf, bois de boeuf'

Bailey

Nev 1867(K — as *sp.* nr. *acuminata*)

Localities recorded: M

Ornamental tree with bilobed leaves to 12 cm long, the veins beneath hairy, flowers with creamy pink petals dotted darker pink to 5 cm long and woody oblong rough pods to 20 cm long.

Brownea Jacq.

B. grandiceps Jacq.   'rose of venezuela'

Bailey, Lionnet (2a)

Localities recorded: M(BG)

Large ornamental tree with pinnate leaves to 17 cm long with up to 11 pairs of leaflets, the young leaves pink in drooping clusters, and dense globose heads of bright red flowers to 20 cm wide, each flower to 8 cm long.

B. latifolia Jacq.   'rose of venezuela'

Lionnet (2),(2a)

Jeff 597(S)

Localities recorded: M(BG)

Ornamental tree similar to *B.grandiceps* but with up to 3 pairs of leaflets and the flower heads up to 12 cm wide.

Caesalpinia L.

C. bonduc (L.)Roxb. (*C. bonducella*)   'cadoque'

Summerhayes, Bailey, Gwynne, Renvoize, Fosberg (1), (2), Stoddart (1),(3)

Rob 2738,3133(S)

Localities recorded: M,F,Co,Bir,Rem,Ald,Ass,Cos,Ast

Indigenous prickly spreading strand shrub with bipinnate leaves, the ovate leaflets to 4.5 cm long and with recurved prickles beneath, erect spikes of greenish yellow flowers, each to 1 cm long, and densely prickly oblong pods to 8 cm long containing grey seeds.

C. pulcherrima (L.)Swartz   'aigrette, poincillade, pride of barbados'

Summerhayes, Bailey, Baumer, Robertson (7)

Jef 748(s), Tod 24(S)

Localities recorded: M,F,Dar

Cultivated shrub with bipinnate leaves, the leaflets oblong and to 2 cm long, terminal pyramidal spikes of long-stalked yellow, orange or pink flowers to 2 cm wide with stamens to 6 cm long, and hard brown oblong pods to 10 cm long which split to fling out the flattened square seeds. Used medicinally in Seychelles.

Cassia L.

C. aldabrensis Hemsley   'cassie'

Renvoize, Fosberg (2)

Localities recorded: Ald,Ass

Endemic herb to 30 cm with wiry reddish brown stems, pinnate leaves with many small leaflets, each to 4 mm long, yellow flowers to 1.2 cm wide and hairy elongated dehiscing pods to 3 cm long.

C. alata L.   'catepan, candle bush, ringworm plant'

Summerhayes, Bailey, Lionnet(1), Baumer

Localities recorded: M(sr only),L

Shrub to 2 m or more with pinnate leaves to 1m long, the oblong leaflets to 19 cm long, erect spikes of showy orange yellow flowers to 2.5 cm wide and straight black pods with a wing down each side to 15 cm long and containing up to 60 flattened triangular seeds. Used medicinally in Seychelles.

C. fistula L.   'butor, caneficier, cassie, indian laburnum'

Bailey

Localities recorded: M(sr only)

Ornamental tree to 10 m with pinnate leaves, the leaflets pointed ovate to 12 cm long, hanging racemes to 60 cm long of attractive pale yellow flowers to 5 cm wide and narrow cylindrical pods to 50 cm long.

C. grandis L.f.   'cassie grandis, horse cassia'

Bailey

No locality recorded

Spreading tree with pinnate leaves, the leaflets oblong and hairy beneath to 5 cm long, short dense clusters of pink and yellow flowers changing to salmon pink and coming before the leaves, each flower to 1.8 cm wide, and slightly curved long dark brown pods to 60 cm long with 2 ridges on one side and one on the other, containing pale brown flat seeds in a smelly pulp.

C. hirsuta L.

Jef 738 (S)

Localities recorded: M

Erect branched pubescent woody herb or shrub with pinnate leaves with up to 12 pairs of ovate leaflets, each to 8 cm long, yellow flowers to 2.5 cm wide in axillary pairs or terminal clusters and long narrow slightly curved and hairy pods to 15 cm long containing small dark olive flattened seeds.

C. mimosoides L.

Summerhayes

Jef 574(S), Rob 2458(S)

Localities recorded: M

Usually erect woody herb with pinnate leaves to 10 cm long with many oblong leaflets each to 5 mm long, yellow flowers to 1.5 cm wide with reddish brown sepals and hairy papery brown pods to 5 cm long, held erect and the valves twisting on dehiscence. (Fig. 41)

C. multijuga Rich.   'cassie jaune'

Bailey

Jef 742(S)

Localities recorded: M

Ornamental shrub or small tree with pinnate leaves, the leaflets oblong and to 3 cm long, sprays of bright yellow flowers each to 2 cm wide and flat thin pods to 15 cm long.

C. nodosa Roxb. 'cassie rose'

Bailey

No locality recorded

Tree to 20 m with pinnate leaves, the ovate pointed leaflets to 9 cm long, dense clusters of sweet-scented deep rose pink flowers, each to 2 cm long, with reddish bracts, fading with age, and long cylindrical pods to 60 cm long with transverse partitions.

C. occidentalis L. 'casse puante, stinking weed'

Summerhayes, Bailey, Gwynne, Renvoize, Fosberg (2), Stoddart (1),(3), Baumer, Wilson (4), Robertson (7)

Jef 708(S), Fos 52069(S), Fra 450(S), Bru/War 33(S), Rob 2366, 2597, 2692,2828,3008,3144(S)

Localities recorded: M,F,N,A,L,Co,Bir,Den,Pla,Coe,Rem,Dar,Poi,Mar,Alp, Pro,Ald,Cos,Ast

Pantropical weed, an erect herb with reddish young stems, pinnate leaves with up to 5 pairs of leaflets, the largest at the top, pointed ovate to 12 cm long, yellow flowers to 2 cm wide in the leaf axils and erect narrow and slightly curved brown pods to 12 cm long with paler margins containing flattened grey brown seeds. Used medicinally in Seychelles. (Fig. 42)

C. polyphylla Jacq.

Rob 3194(S)

Localities recorded: M

Ornamental shrub to 2 m with clusters of pinnate leaves to 5 cm long with up to 15 pairs of small obovate leaflets each to 7 mm long which fall early leaving the rachis, bright yellow flowers to 3.5 cm wide in pairs among the leaves and small elongated flattened papery dark brown pods to 10 cm long with indentations between the seeds.

C. sophera L.

Jef 393(S)

Localities recorded: M

Shrub to 2 m with pinnate leaves to 25 cm long or more with up to 10 pairs of lanceolate leaflets each to 7 cm long, yellow flowers to 3 cm wide from the leaf axils and straight or slightly curved subcylindrical pods to 10 cm long. Similar to *C. occidentalis* but with more leaflets.

Fig. 41. Cassia mimosoides

Fig. 42. Cassia occidentalis

Fig. 43. Hymenaea verrucosa

Fig. 44. Intsia bijuga

C. tora L.   'casse puante petite espece'

Bailey

Rob 3185 (S)

Localities recorded: An

Erect herb, a weed, similar to *C. occidentalis* and *C. sophera* but with 3 pairs of rounded leaflets, small yellow flowers to 1 cm wide in the leaf axils and narrow curved pods to 15 cm long, only slightly margined and twisting spirally when splitting to show the cylindrical brown seeds.

Cynometra L.

C. cauliflora L.   'nam nam'

Lionnet (2)

Rob 2754(S)

Localities recorded: M(BG)

Cultivated small tree with leaves of 2 unequal-sided leaflets to 10 cm long, small pink flowers in clusters on the trunk and main branches and pulpy flat pods to 10 cm long, edible but acid tasting, containing one large seed.

Delonix Raf.

D. regia (Hook.)Raf. (*Poinciana regia*)   'flamboyant, flame tree'

Bailey, Lionnet (1),(2), Procter (2), Fosberg (2), Stoddart (1), Robertson (4),(5)

Rob 3126(S)

Localities recorded: M(sr only),P,F,N,Coe,Des,Ald

Ornamental spreading tree with bipinnate leaves with many oblong leaflets, each to 1.5 cm long, large axillary clusters of showy red and orange flowers, each to 12 cm wide, and hard long flattened brown pods to 40 cm long containing oblong pale brown and purplish mottled seeds, each to 2.5 cm long.

Haematoxylon L.

H. campechianum L.   'bois campeche, logwood'

Summerhayes, Bailey

Pro 3940(S)

Localities recorded: M,D

Small slow-growing spiny tree with odd-pinnate leaves, the leaflets heart-shaped and to 4 cm long, axillary sprays of yellow scented flowers to 1cm wide and flattened ovate pods to 5 cm long, opening down the middle of the valves. The heart wood and roots yield a purple dye.

Hymenaea L.

H. verrucosa Gaertn. (*Trachylobium verrucosum*)   'copalier, gum copal'

Summerhayes, Bailey

Pro 4147(S)

Localities recorded: M,Ro

Cultivated and escaping tree to 20 m with leaves with one pair of unequal-sided leathery curving leaflets, each to 12 cm long, axillary clusters of pink and white flowers to 2 cm long and ovoid fruit to 4 cm long covered in irregular gum warts and containing one or two seeds. (Fig. 43)

Intsia Thou.

I. bijuga (Colebr.)O.Kuntze   'gayac'

Summerhayes, Bailey, Lionnet (1), Procter (2), Wilson (2)

Jef 617(S), Rob 2517(S), War 196(S)

Localities recorded: M,P,C,A,Fe,T

Indigenous tree with pale grey bark, pinnate leaves with 2 pairs of ovate leaflets, each to 15 cm long and curving forwards, flat clusters of flowers, each with one pink petal and red stamens, and oblong flat pods to 25 cm long with a few flat round seeds. (Fig. 44)

Lysidice Hance

L. rhodostegia Hance   'casse'

Dup 4/31,4/33,20/33(K)

Localities recorded: M(BG)

Cultivated small tree with pinnate leaves with 4 pairs of long ovate leaflets, the larger to 14 cm long and curving forwards, showy terminal sprays of violet flowers to 2 cm wide with 2.5 cm long rose bracts and flat oblong woody pods to 15 cm long.

Peltophorum (Vogel)Walp.

P. pterocarpum (DC.)K.Heyne (*P.ferrugineum, P.inerme*)   '?calice'

Bailey, Lionnet (2), Procter (2)

Localities recorded: M(BG),P

Large ornamental tree with spreading crown, feathery bipinnate leaves with many oblong leaflets each to 1.5 cm long, erect sprays of showy yellow flowers, rusty brown in bud, each to 3 cm wide, and reddish brown irregularly oblong flattened pods to 10 cm long containing one or two seeds.

Saraca L.

S. indica L   'jonesia asoka'

Bailey

Jef 605(S)

Localities recorded: M(BG)

Small spreading tree with pinnate leaves with up to 6 pairs of oblong leaflets, each to 20 cm long, young leaves in greyish white drooping clusters, large sessile clusters of orange scented, apparently tubular, flowers, each to 3 cm long.

Tamarindus L.

T.indica L.   'tamarinier, tamare, tamarind'

Bailey, Gwynne, Fosberg (2), Stoddart (1), Baumer, Wilson (1),(2), Robertson (4),(5)

Jef 755(S), Rob 3150(S)

Localities recorded: M,P,F,N,C,Coe,Des,Ald

Cultivated and escaping large spreading tree with pinnate leaves with up to 18 pairs of oblong leaflets, each to 3 cm long, yellow flowers veined with red to 2 cm long and irregular brown swollen pods to 14 cm long containing a brown sweet acid pulp and several squarish seeds. The pulp is used to flavour food and the plant is used medicinally in Seychelles. (Fig. 45)

LEGUMINOSAE: MIMOSOIDEAE (57)

Acacia Mill.

A. confusa Merr.   'acacia petit feuille, mimosa petit feuille'

Summerhayes, Bailey, Wilson (2)

Dup 1/30(K), Jef 414(S), Pro 4172,4282(S)

Localities recorded: M,F,C

Tree to 15 m with curved lanceolate phyllodes to 10 cm long, shortly stalked axillary globose heads of yellow flowers, the heads to 8 mm wide, and flat pods to 10 cm long, with square segments.

A. farnesiana (L.)Willd.   'fragrant acacia'

Summerhayes

Localities recorded: M

Spiny shrub or tree to 4 m with feathery bipinnate leaves, the leaflets to 7 mm long, and small golden heads, to 1 cm wide, of fragrant flowers, with short fat sometimes curved dark brown pods to 7.5 cm long.

A. mearnsii De Wild.   'black wattle'

Bailey

No locality recorded

Tree with rounded crown, pubescent twigs and feathery bipinnate leaves, the leaflets to 4 mm long, fragrant pale yellow flowers in heads to 8 mm wide and narrow curved flat pods to 10 cm long with hollows between the seeds.

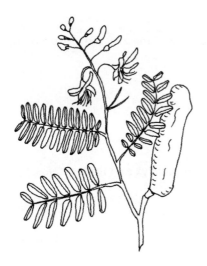

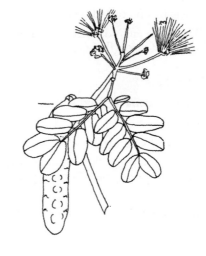

Fig. 45. Tamarindus indica   Fig. 46. Albizia lebbeck

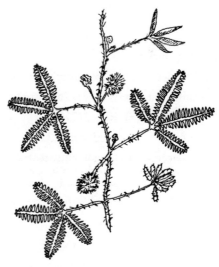

Fig. 47. Leucaena leucocephala   Fig. 48. Mimosa pudica

A. polystachya Mill.   'acacie grande feuille, wattle'

Bailey

Dup 3/31(K)

Localities recorded: ?M

Tree with curved lanceolate phyllodes to 18 cm long and slender branched sprays of small flowers.

Adenanthera L.

A. microsperma Teijsm. & Binn.   'agati petite feuille'

Bailey

No locality recorded

Tree similar to *A. pavonina* but with pubescent calyx and flower stems, pods twisting spirally before they split and smaller seeds.

A. pavonina L.   'agati, bead tree, red sandalwood'

Summerhayes, Sumner Gibson, Bailey, Gwynne, Lionnet (1), Renvoize, Stoddart (2), Wilson (1),(2), Robertson (3),(4)

Jef 370,695,729(S), Fos 52162(S), Rob 2841,3123(S)

Localities recorded: M,S,P,F,C,Co,Coe,Dar,Poi,Alp

Possibly indigenous tree with bipinnate leaves with ovate to rounded leaflets, each to 4 cm long, spikes to 20 cm long of small scented brownish yellow flowers and clusters of long curved pods to 30 cm long which twist open to show a golden interior and the bright red flattish round seeds, each to 1 cm wide.

Albizia Durazz.

A. falcata (L.)Back. ex Merr. (*A.moluccana*)   'albizzia'

Bailey, Lionnet (1),(2),(2a), Procter (2), Wilson (2), Robertson (4)

Jef 405,739(S), Rob 2471(S)

Localities recorded: M,P,F,C

Tall tree with greyish white trunk, large feathery bipinnate leaves, the leaflets oblong and pointed and to 2 cm long, the branches often horizontal giving the tree a flat-topped look, clusters of spikes of small white flowers, each to 1 cm long, and flat oblong dark brown pods to 10 cm long containing small flat seeds.

A. lebbeck (L.)Benth.   'bois noir, woman's tongue'

Summerhayes, Bailey, Lionnet (1), Wilson (1), Robertson (3),(6)

Jef 1140(S), Rob 2720(S)

Localities recorded: M,S,P,F,An,Poi

Tall timber tree with bipinnate leaves with rounded oblong leaflets, each to 5 cm long, globose heads of greenish yellow long-stamened flowers, each to 3 cm long, and flattened pale yellowish brown pods to 33 cm long with prominent veining which rattle in the wind when ripe. (Fig. 46)

Calliandra Benth.

C. alternans Vahl ex Benth.

Renvoize, Fosberg (2)

Localities recorded: Ald

Shrubby tree to 5 m with bipinnate leaves with one pair of pinnae, the leaflets oblong and to 12 mm long, globose heads of yellow orange long stamened flowers, each to 1.5 cm long, and narrow pods which taper to the base to 16 cm long.

C. surinamensis Benth.   'calliandra'

Localities recorded: M(sr only)

Low spreading ornamental shrub with an inverted cone shape with feathery bipinnate leaves, the leaflets to 1.5 cm long, brush-like heads of reddish white long-stamened flowers, each to 4 cm long, and thick edged narrow pods to 10 cm long.

Desmanthus Willd.

D. virgatus (L.)Willd.

Summerhayes, Renvoize, Stoddart (1), Robertson (6)

Jef 707(S), Pro 4131(S), Rob 2753,2961,3127(S)

Localities recorded: M,An,Coe,Dar,Far

Slender, sometimes scandent, shrub with feathery bipinnate leaves, the leaflets narrowly oblong and to 1 cm long, loose heads of small white flowers, each to 1cm long, and narrow flat brown pods to 10cm long and 4 mm wide.

Dichrostachys (DC.)Wight & Arn.

D.microcephala Renvoize   'trefle'

Renvoize, Fosberg (2)

Localities recorded: Ald,Cos,Ast

Tree or shrub to 4 m with feathery bipinnate leaves with oblong leaflets to 1.5 mm long, flowers in small erect oblong spikes to 1 cm long, the upper flowers pale yellow and the lower pale pink, and small solitary flattened oblong pods to 2.5 cm long.

Entada Adans.

E. ?pursaetha DC.

Localities recorded: P(sr only)

A seed washed up on the beach on Praslin was planted in a pot and germinated but there is so far no record of a naturally occurring plant. It is a liane with bipinnate leaves, the leaflets oblong to 9 cm long, with spikes to 23 cm long of small creamy yellow flowers with gigantic pods to 2 m long and 15 cm wide containing hard flattened squarish dark brown seeds to 5 cm wide.

Leucaena Benth.

L. leucocephala (Lam.)de Wit (*L.glauca*)   'cassie, wild tamarind'

Summerhayes, Gwynne, Renvoize, Stoddart (1),(2), Wilson (1),(2),(4), Robertson (3),(4),(6)

Jef 671(S), Pra 445(S), Rob 2299,2791,3005,3132(S)

Localities recorded: M,S,P,F,C,An,Pla,Coe,Rem,Dar,Jos, Mar,Far

Rapidly growing shrub with bipinnate leaves, the leaflets oblong and pointed to 1.5 cm long, globose heads to 2.5 cm wide of small white flowers and clusters of elongated flat brown pods, each to 20 cm long, containing small flat ovate dark brown shiny seeds. The foliage is used as forage. (Fig.47)

Mimosa L.

M. pudica L.   'herbe sensible, sensitive plants, shame plant'

Summerhayes, Bailey, Lionnet (1), Baumer, Wilson (1), Robertson (6)

Sch 11707(K), Rob 2487,2698,3158(S)

Localities recorded: M,P,F,An,Coe

Prostrate spiny spreading hairy herb with bipinnate leaves, the leaflets narrow to 1 cm long, with two pairs of pinnae which close up when touched, globose heads to 1.5 cm wide of pink flowers and flattened prickly pods to 2 cm long, the segments falling out when ripe leaving the spiny sutures. Used medicinally in Seychelles. (Fig. 48)

Parkia R.Br.

P. pedunculata (Roxb.)Macbr. (*P. biglandulosa*) 'kiboko'

Bailey(probably his *P. roxburghii*, now called *P. javanica*), Lionnet (2)

Localities recorded: M(BG)

Cultivated tall tree with bipinnate leaves with up to 100 pairs of leaflets, flowers in dense heads, the upper part globose then narrowing to a neck, on long stalks, white at the base and tawny above, and clusters of brown pods to 45 cm long containing a white powdery substance.

Pithecellobium Mart.

P. dulce (Roxb.)Benth.   'madras thorn'

Rob 3174(S)

Localities recorded: M

Spiny tree to 15 m with bipinnate leaves of 4 obovate leaflets, each to 5 cm long, tight clusters of whitish flowers, each to 8 mm long, on long spikes and twisted pale brown or pinkish pods constricted between the seeds which are up to 1 cm long and black with a white or red aril. (Fig. 49)

P. unguis-cati (L.)Benth. 'campeche'

Bailey, Stoddart (2), Gwynne, Wilson (2), Robertson (3),(6)

Pro 4127(S), Tod 49(S), Rob 3143(S)

Localities recorded: M,F,C,An,Coe,Jos,Poi

Spiny shrub with bipinnate leaves of 4 rounded to obovate leaflets, each to 5 cm long, long-stalked loosely globose heads of greenish white flowers with many stamens, each to 1 cm long, and dark reddish brown pods which twist when ripe and split to show the squarish black seeds with a white aril which turns bright red.

Samanea (Benth.)Merr.

S. saman (Jacq.)Merr. (*Pithecellobium saman*) 'saman, rain tree'

Bailey, Wilson (1), Robertson (6)

Localities recorded:: P,An

Large tree with short trunk and widely spreading branches, bipinnate leaves with obovate pubescent leaflets to 7 cm long, stalked heads of greenish white flowers with long pink stamens, each flower to 5 cm long, and elongated ridge-edged pods to 20 cm long containing seeds embedded in pulp.

LEGUMINOSAE: PAPILIONOIDEAE (57)

Abrus L.

A. precatorius L. 'liane reglisse, reglisse, indian liquorice, crab's eyes, rosary pea'

Summerhayes, Bailey, Lionnet (1), Renvoize, Fosberg (2), Wilson (1),(2), Robertson (6)

Jef 413(K),531(S), Bru/War 58(S), Rob 2588,2680, 3157(S)

Localities recorded: M,S,P,F,N,C,A,An,Coe,Ald,Ass

Slender twining vine with pinnate leaves with 12 pairs or more of oblong leaflets to 1.5 cm long or more, short axillary clusters of pink or purple flowers, each to 1 cm long, and oblong pods to 4 cm long which split to show the oval red and black seeds which are poisonous. (Fig. 50)

Alysicarpus Desv.

A. vaginalis (L.)DC.

Summerhayes

Dup 1/31(K), Jef 717(S)

Localities recorded: M

Spreading herb, woody at the base, with rounded ovate leaves to 2 cm long, spikes of orangey pink or purple flowers, each to 6 mm long and cylindrical pods to 1.5 cm long which break up transversely into sections when ripe.

Arachis L.

A. hypogea L.   'pistache, groundnut, peanut'

Bailey

Localities recorded: M(sr only)

Cultivated erect or trailing herb with pinnate leaves with two pairs of obovate leaflets, each to 4 cm or more long, yellow flowers to 1.5 cm long and pods to 6 cm long which bury themselves underground as they are developing and contain up to two reddish brown skinned edible nuts.

Cajanus DC.

C. cajan (L.)Millsp.   'ambrevade, pigeon pea, red gram'

Summerhayes, Bailey, Baumer

No locality recorded

Cultivated erect woody herb or shrub to 4 m with trifoliolate leaves, the elliptic leaflets to 9 cm long, yellow flowers to 2 cm long with red or purple markings and pods flattened between the seeds. Used as a green bean and as a pulse and used medicinally in Seychelles.

Calopogonium Desv.

C. mucunoides Desv.

Rob 2710(S)

Localities recorded: F

Perennial trailer covered with rusty hairs, with trifoliolate leaves, the leaflets rounded to ovate to 10 cm long, lax clusters of small blue flowers, each to 7 mm long and flattened hairy pods to 4 cm long containing brown or yellowish seeds.

Canavalia DC.

C. cathartica Thouars (*C.microcarpa*)

Summmerhayes, Gwynne, Wilson (2), Robertson (7)

Jef 651(S), Fos 52138(S), Bru/War 22(S), Rob 2287, 2649,3060(S)

Localities recorded: M,F,N,C,A,Co,Coe,Poi

Indigenous strand plant, a vigorous climber with trifoliolate leaves, the leaflets ovate to 10 cm long, hanging clusters of scented pink or mauve flowers, each to 2 cm long, and pale brown oblong pods to 12 cm long with two ridges at one edge containing smooth reddish brown seeds with stalk-scar to 14 mm long.

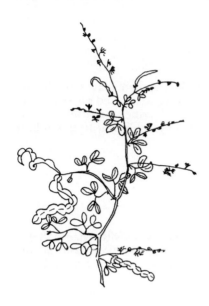

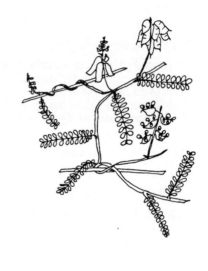

Fig. 49. Pithecellobium dulce

Fig. 50. Abrus precatorius

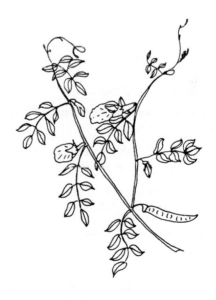

Fig. 51. Centrosema pubescens

Fig. 52. Clitoria ternatea

63

C. ensiformis (L.)DC.   'dolique, gros pois, pois sabre (but see *C.gladiata*), jack bean, horse bean'

Bailey

Rob 3198(S)

Localities recorded: M

Cultivated plant with scandent branches, and trifoliolate leaves, the leaflets ovate to 20 cm long, arching spikes of mauve flowers streaked with white and pods to 35 cm long with two ridges down one side, ripening straw coloured and containing white seeds to 3 cm long with the stalk-scar to 9 mm long. Young pods used as a vegetable.

C. gladiata DC.   'sword bean'

Summerhayes, Stoddart (3)

Localities recorded: M,Den

Cultivated large perennial climber with trifoliolate leaves with ovate leaflets to 20 cm long, spikes of white or pinkish flowers and curved pods to 40 cm long with strongly developed ridges and containing reddish seeds with a stalk-scar to 2 cm long. The young pods used as a vegetable.

C. rosea (Sw.)DC. (*C.obtusifolia, C.maritima*)

Summerhayes, Renvoize, Fosberg (2)

Jef 1184(S)

Localities recorded: F,Ald,Ass

Indigenous strand plant, a climber with trifoliolate leaves, the leaflets ovate and to 9 cm long, pinkish mauve flowers and oblong curved pods to 15 cm long with a double ridge on the concave side and darkly mottled brown seeds with the stalk-scar to 8.5 mm long.

Centrosema (DC.)Benth.

C. pubescens Benth.   'centrosema petite feuille, centro'

Bailey, Swabey

Rob 2431,2744(S), Tod 31(S)

Localities recorded: M,F,Fe

Slender trailing pubescent perennial with trifoliolate leaves, the leaflets ovate and to 7 cm long, few-flowered clusters of large pale mauve and white flowers, each to 2.5 cm wide, and long thin flattened pods to 17 cm long with ribbed edges containing small dark brown seeds. (Fig. 51)

Clitoria L.

C. ternatea L.   'liane madame, liane ternate'

Summerhayes, Bailey, Renvoize, Fosberg (2)

Localities recorded: M,Ald

Slender trailing vine with odd-pinnate leaves with up to 7 elliptic or rounded leaflets, each to 4 cm long or more, solitary or paired flowers, white, pale or bright blue and to 4.5 cm wide, and narrow flat pods to 10 cm long containing mottled seeds. (Fig. 52)

Crotalaria L.

C. berteroana DC. (*C. fulva*)

Summerhayes

Rob 2460(S)

Localities recorded: M

Indigenous herb, woody at base, with simple oblanceolate pubescent leaves to 10 cm long, yellow flowers to 1 cm long in dense terminal leafy spikes and inflated pods containing kidney shaped seeds.

C. juncea L.   'cascavelle, sunn hemp'

Bailey

No locality recorded

Erect shrubby hairy annual to 2 m with simple lanceolate leaves to 13 cm long, showy yellow flowers to 2 cm long in terminal spikes and cylindrical hairy light brown pods to 4 cm long containing several dark grey or black kidney shaped seeds to 6 mm long.

C. laburnifolia L.

Summerhayes

Localities recorded: M

Erect herb with trifoliolate leaves, the leaflets obovate and to 10 cm long, yellow flowers to 2 cm long often marked with red or brown stripes or flushes, and inflated pods to 15 cm long, mottled red or brown and with a long stipe (the narrow base of the pod), containing brown seeds to 5 mm long.

C. laburnoides Klotsch

Renvoize, Fosberg (2)

Localities recorded: Ald,Cos

Erect herb to 60 cm with trifoliolate leaves, the leaflets ovate to 4 cm long, yellow flowers to 1 cm long clustered in terminal spikes, and short inflated pubescent pods to 2 cm long containing orange brown heart shaped seeds to 2.5 cm long.

C. pallida Ait. (*C.mucronata, C.striata*)   'cascavelle trois feuilles'

Bailey

Jef 751(S), Lam 404(K), Pro 4133,4160(S), Rob 2235, 2778,3188(S), Tod 16(S)

Localities recorded: M,F,C,An

Erect herb to 2 m tall, woody at the base, with trifoliolate leaves, the leaflets ovate to obovate to 10 cm long, spikes of closely arranged yellow flowers, each to 1.5 cm long and veined reddish brown, and small cylindrical slightly curved pods to 4.5 cm long containing many small mottled heart-shaped seeds.

C. retusa L.   'shack shack'

Summerhayes

Jef 776(S), Rob 2440,2684(S)

Localities recorded: M,S,F,L

Indigenous erect annual to 1.3 m tall with simple oblanceolate leaves to 10 cm long, erect terminal spikes of spaced-out yellow flowers to 1.5 cm long and veined with red with the large standard petal red outside, and cylindrical pods to 5 cm long, on stalks (not stipes) containing two rows of large flat seeds. (Fig.53)

C.verrucosa L.   'cascavelle bleu'

Bailey, Stoddart (3)

Localities recorded: Dar

Erect much branched herb to 1 m with widely ovate simple leaves to 12 cm long, long spikes of closely arranged pale yellow flowers tinged with blue, each to 1.3 cm long, and shortly hairy pods to 5 cm long containing shiny brown heart-shaped seeds, each seed with one end incurved.

C. zanzibarica Benth.   'cascavelle trois feuille'

Summerhayes (Horne 484), Bailey

Localities recorded: M

Erect herb with trifoliolate leaves, the leaflets elliptic and to 10 cm long, long stalked spikes of yellow flowers, each to 1.5 cm long and veined reddish purple, and inflated pods to 4.5 cm long, deeply keeled above and containing many smooth orangey buff seeds.

Derris Lour.

D. trifoliata Lour. (*D.uliginosa*)   'derris, tuba'

Summerhayes, Bailey

Pro 4376(S), Rob 2488(S)

Localities recorded: M,S,D

Indigenous woody vine with lenticels on the stem, odd pinnate dark green glossy leaves, with up to 5 leaflets each to 12 cm long, long sprays of pale pink flowers, each to 1 cm long, and flat widely oblong brown veined pods to 4 cm long containing 2 seeds. (Fig. 54)

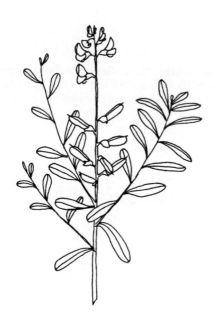

Fig. 53. Crotalaria retusa

Fig. 54. Derris trifoliata

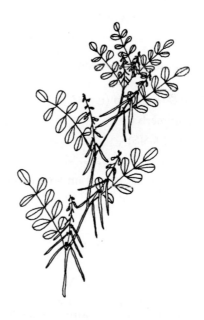

Fig. 55. Flemingia strobilifera

Fig. 56. Indigofera tinctoria

Desmodium Desv.

D. adscendens (Sw.)DC.   'matite'

Summerhayes, Bailey

Localities recorded: M

Herb with creeping stems and trifoliolate leaves, the leaflets rounded and the largest to 2.2 cm long, pink flowers and pods with up to 5 segments. (The only specimen, Horne 287, may not be *D. adscendens* and the above description was taken from FTEA)

D. canum (J.F.Gmel.)Schinz & Thell. (*D. frutescens*, *D. incanum*, *Meibomia cajanifolia*)   'trefle petit'

Summerhayes, Bailey, Sauer, Baumer, Robertson (6), (7)

Dup 5/31(K), Ves 5610(K), Jef 476(K),498(S), Rob 2367, 2575, 2428(S), Tod 48(S)

Localities recorded: M,S,P,F,N,C,Fe,An,L,Poi

Herb, woody at the base, with trifoliolate leaves, the leaflets ovate and the largest to 7 cm long, spikes of widely spaced pink or purplish flowers, each to 5 mm long, and flat sticky pods with up to 8 D-shaped segments which break up easily and stick to clothing. Used medicinally in Seychelles.

D. gangeticum (L.)DC.

Sch 11654(K)

Localities recorded: M

Herb, woody at the base, with simple leaves, white, pink or purple flowers and pods with up to 8 segments, curved. This species is very similar to *D. canum* and may be the same.

D. heterocarpon (L.)DC.

Summerhayes

Pro 4063(S)

Localities recorded: M

Indigenous strand undershrub with trifoliolate leaves, the leaflets obovate and the largest to 6 cm long, dense sprays of small pinkish mauve flowers, each to 5 mm long, and hairy pods with up to 6 segments, each to 5 mm long.

D. ramosissimum G.Don

Jef 704(S)

Localities recorded: M

Woody herb to 1.5 m with trifoliolate leaves, the leaflets obovate and the largest to 4.5 cm long, somewhat sensitive, terminal sprays of reddish or purplish flowers, each to 4 mm long, and pods to 2 cm long with few segments.

D. triflorum (L.)DC. (*D. heterophyllum*)   '?trefle petit, ?tourlas'

Summerhayes, Bailey, Sauer, Robertson (6)

Jef 379(S), Sch 11751(K), Pro 4057(K), Rob 2553,2660, 3156(S)

Localities recorded: M,P,F,N,C,Fe,An,L,Coe

Indigenous slender prostrate hairy herb with trifoliolate leaves, the leaflets heart-shaped and to 1.2 cm long, small purplish flowers, one or two in the leaf axils, and small pods to 1.5 cm long with up to 6 segments. Often in lawns.

D. triquetrum (L.)DC. (*D. auriculatum, Pteroloma auriculatum*)   'trefle gros'

Summerhayes, Bailey, Evans

Ves 6418 (K), Rob 2425(S)

Localities recorded: M,P

Indigenous erect branched woody herb or shrub with simple ovate pointed leaves to 10 cm long with winged petioles, long spikes of purplish flowers, each to 8 mm long, and pods with up to 8 squarish flattened golden segments.

D. umbellatum (L.)DC. (*Dendrolobium umbellatum*)

Summerhayes

Rob 2392(S)

Localities recorded: P,C

Indigenous strand shrub or small tree with trifoliolate leaves, the leaflets ovate and pale beneath to 8 cm long, dense axillary sprays of white or yellow flowers, each to 1 cm long, and hard pods with up to 6 D shaped segments each to 1 cm long.

Dipteryx Schreb.

D. odorata (Aubl.)Willd.   'tonquin, tonka bean'

Bailey

No locality recorded

Tree with pinnate leaves with up to 6 oblong leaflets, each to 20 cm long, terminal sprays of rose violet flowers, each to 1.5 cm long, and pale yellow brown fibrous pods to 4 cm long containing a single dark brown fragrant seed.

Erythrina L.

E. corallodendrum L.   'mourouc a fleur de corail, coral bean tree'

Bailey

No locality recorded

Ornamental small tree with trifoliolate leaves, the leaflets ovate to 12 cm wide, spikes of red flowers, each flower with the largest petal to 5 cm long, and pods constricted between the seeds which are scarlet with a black spot.

E. crista-galli L.

Lionnet (2)

Localities recorded: M(BG)

Cultivated tree with trifoliolate leaves, the terminal leaflet to 7cm long, and solitary or few-flowered terminal clusters of red flowers, the large petal to 5cm long, and woody pods to 35cm long, constricted between the blackish seeds.

E. variegata L.   'mourouc or nourouc'

Summerhayes, Bailey, Fosberg (2), Baumer

Rob 2838(S)

Localities recorded: M,P,Alp,Ald

Possibly indigenous strand tree to 12 m, or an ornamental, with prickles on the branches, trifoliolate leaves, the leaflets triangular and the lateral ones unequal, to 22 cm long, dense clusters of scarlet flowers, each to 6 cm long and usually before the leaves, and pods to 30 cm long, constricted between the seeds which can number up to 14 and are pink and oblong. Used medicinally in Seychelles.

Flemingia Ait.f.

F. strobilifera (L.)Ait.f. (*Moghania strobilifera*)   'napoleon'

Summerhayes, Bailey, Lionnet (1)

Jef 535(S), Sch 11680(K), Rob 2456(S)

Localities recorded: M,L

Velvety erect branched shrub with ovate leaves to 15 cm long, striking sprays on which the small flowers and swollen pods, each to 1 cm long, are enclosed in large green overlapping bracts, each to 3 cm long, which turn pale golden brown and papery. (Fig. 55)

Gliricidia Kunth.

G. sepium (Jacq.)Walp.   'gliricidia'

Bailey, Gwynne, Swabey

Sch 11721(K)

Localities recorded: M,P,Coe

Fast growing ornamental tree with odd-pinnate leaves to 30 cm long, the leaflets ovate pointed, dense axillary clusters of pink flowers, each to 2 cm long, and narrowly oblong flat pods with up to 8 seeds. Used for shade or for vanilla supports.

Indigofera L.

I. spicata Forsk.   'indigo'

Bailey

No locality recorded

Woody herb with ridged stems, odd-pinnate leaves with up to 11 alternate leaflets, each to 1 cm or more long, dense clusters of small pink flowers, each to 5 mm long, and downward pointing straight pods to 1.8 cm long, containing smooth yellow seeds.

I. suffruticosa Mill. 'indigo'

Summerhayes

Jef 839(S), Pro 4173(S), Rob 2450,2459,2631,2694(S)

Localities recorded: M,S,F,N

Erect silvery pubescent herb, woody at base, with odd-pinnate leaves with up to 13 obovate leaflets, each to 3 cm long, axillary spikes of many small pink flowers, each to 4 mm long, and cylindrical pubescent curved pods to 1.5 cm long containing brown oblong seeds.

I. tinctoria L. 'indigo sauvage'

Bailey, Baumer

Jef 623(S), Pro 4051(S), Rob 3165,3186(S)

Localities recorded: M,Fe,An

Shrubby branched herb to 1.7 m tall with odd-pinnate leaves with up to 13 obovate leaflets, each to 2 cm long, axillary spikes of many small orange yellow flowers, each to 6 mm long, and greyish brown straightish downward pointing pods to 3.5 cm long and containing dark brown oblong seeds. Used medicinally in Seychelles. (Fig. 56)

I. sp.

Fosberg (2) quotes an *Indigofera* on Aldabra and Stoddart (1) quotes one on Farquahar.

Lablab Adans.

L. purpureus (L.)Sweet (*Dolichos lablab*) 'lablab'

Summerhayes

Localities recorded: M

Climbing perennial bean with trifoliolate leaves, the leaflets ovate triangular and to 15 cm long, purple and white flowers to 1.5 cm long in few flowered clusters on long stiff sprays and broad pods to 14 cm long containing up to 4 seeds.

Macroptilium (Benth.)Urb.

M. atropurpureum (DC.)Urb. cv. siratro (*Phaseolus atropurpureus*) 'siratro'

Rob 2997(S)

Localities recorded: Dar

Cultivated fodder crop, a trailing perennial with trifoliolate leaves with rhomboid leaflets to 5 cm long, and long-stalked few-flowered clusters of deep purple flowers.

Mucuna Adans.

M. gigantea (Willd.)DC.   'liane caiman, liane cadoque'

Summerhayes

Ves 5772(K)

Localities recorded: M,S

Indigenous strand plant, a large woody liane with trifoliolate leaves, the leaflets ovate with laterals oblique to 15 cm long, young stems and calyces with stiff orange hairs, clusters of greenish white or lilac flowers, each to 4 cm long and turning black, and flattened oblong pods to 12 cm long and 5 cm wide covered with irritant orange bristles and containing discoid black mottled brown seeds

Phaseolus L.

P. lunatus L.   'pois du cap, lima bean'

Summerhayes, Bailey

Rob 2985(S)

Localities recorded: S,Dar

Twining perennial vegetable, rarely bushy, with trifoliolate leaves, axillary sprays of many pale green and white flowers and oblong recurved pods containing variable seeds which have lines radiating from the white stalk-scar to the edges.

P. vulgaris L.   'haricot, common, kidney, french or runner bean'

Bailey

Localities recorded: M(sr only)

Twining or erect annual vegetable with many cultivated forms, having trifoliolate leaves, axillary few flowered lax clusters of white, yellow, pink or violet flowers, long slender pods variable in colour when ripe and containing usually oblong seeds with a white stalk-scar.

Pterocarpus Jacq.

P. indicus Willd.   'sang dragon, bloodwood'

Bailey, Lionnet(1), Wilson (1),(2), Robertson (6)

Rob 2290, 2737(S)

Localities recorded: M,P,F,C,An

Large graceful tree to 30 m with buttresses and drooping branches of odd-pinnate leaves, the leaflets alternate, ovate pointed to 10 cm long, axillary clusters of many small yellow scented flowers, each to 1 cm long and quickly falling, and flat circular winged pods to 6 cm wide. (Fig. 57)

Pueraria DC.

P. phaseoloides (Roxb.)Benth.   'centro grande feuille, puero, tropical kudzu'

Bailey

Rob 3169(S)

Localities recorded: M

Trailing or climbing hairy vine, a pasture legume, with trifoliolate leaves, the leaflets rhomboid to 10 cm long, mauve flowers to 1.2 cm long in clusters on erect hairy stalks and flat straight hairy pods to 10 cm long containing oblong rounded red seeds.

Sesbania Adans.

S. bispinosa (Jacq.)W.F.Wight (*S.aculeata*)

Summerhayes

Dup 6/33(K), Fos 52051(S), Rob 2393(S)

Localities recorded: M,Co

Slender slightly woody herb to 3 m with pinnate leaves to 35 cm long with up to 55 pairs of oblong leaflets, each to 2 cm long, slender few-flowered sprays of brownish yellow flowers, each to 1.5 cm long and slender cylindrical pods to 25 cm long containing brown seeds in separate compartments.

S. grandiflora (L.)Poir.   'agati, brede malabar, brede morongue gros'

Bailey

No locality recorded

Small ornamental tree with pinnate leaves to 30 cm long with obovate leaflets to 3 cm long, short axillary sprays of a few large white flowers, each to 10 cm long, and narrow slightly curved pods to 40 cm long containing many seeds.

S. sericea (Willd.)Link

Summerhayes, Gwynne, Renvoize, Stoddart (2),(3), Wilson (3),(4)

Bru/War 104(S), Rob 2832,2943(S)

Localities recorded: A,Bir,Den,Dar,Mar?,Def?,Alp

Woody herb to 3 m, pubescent on stems and leaves, with pinnate leaves to 15 cm long with up to 25 pairs of leaflets, each to 2 cm long, small creamy flowers flecked with purple to 1cm long and slender cylindrical pods to 15cm long with flecked brown seeds in separate compartments.

Sophora L.

S. tomentosa L.   'bois chapelet'

73

Renvoize, Fosberg (2)

Horne 281(K), Veevers Carter 95(K), Jef 843(S), Rob 2726,3166(S)

Localities reorded: M,F,Pie,Ald,Ass,Ast

Probably indigenous strand plant, a shrub to 3 m with grey pubescent odd-pinnate leaves, the leaflets broadly ovate to 4 cm long, dense terminal sprays of yellow flowers, each to 2 cm long, and hanging pods to 20 cm long, constricted between the seeds which are round hard-shelled and dark brown. (Fig.58)

Tephrosia Pers.

T. noctiflora Bak.

Summerhayes, Wilson (2)

Jef 1176(S), Pro 4159(S), Rob 2673,3170(S)

Localities recorded: M,F,C

Indigenous softly hairy woody herb with odd-pinnate leaves, with up to 25 oblanceolate leaflets, each to 4 cm long and parallel veined beneath, terminal spikes of widely spaced purplish white flowers, each to 1cm long, and flattened oblong softly hairy orange pods to 6 cm long containing well spaced small dark kidney-shaped seeds.

T. pumila (Lam.)Pers. var. **aldabrensis** (J.R.Drummond & Hemsley) Brummitt 'indigo sauvage'

Renvoize, Fosberg (2)

Localities recorded: Ald,Ass,Cos,Ast

Prostrate or erect herb with odd pinnate leaves with up to 11 narrow leaflets to 8 mm long and hairy below with parallel veins, solitary or few mauve, pink or white flowers to 8 mm long on terminal clusters and linear hairy pods to 3 cm long containing up to 12 brown flattened seeds.

T. purpurea (L.)Pers.   'tephrosia'

Summerhayes, Bailey

No locality recorded

Erect or spreading woody herb with slender odd pinnate leaves with up to 17 obovate leaflets each to 2.5 cm long and with few parallel veins beneath, slender spikes of red or pink flowers, each to 8 mm long, and narrow pods to 4 cm long, upturned at the tips and containing up to 9 widely spaced small mottled cylindrical seeds. (Fig. 59)

Teramnus P.Br.

T. labialis (L.f.)Spreng.

Summerhayes, Renvoize, Fosberg(2), Wilson (2)

Lionnet 1(K), Pro 4053(S), Rob 2573,2647,3189(S)

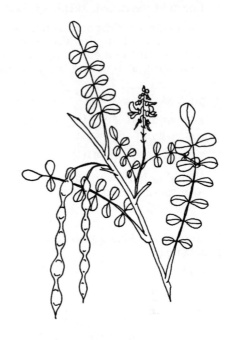

Fig. 57. Pterocarpus indicus

Fig. 58. Sophora tomentosa

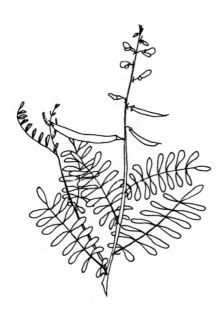

Fig. 59. Tephrosia purpurea

Fig. 60. Teramnus labialis

Localities recorded: M,D,F,N,C,Fe,An,Ald

Indigenous trailing or prostrate herb with trifoliolate leaves, the leaflets ovate and to 7 cm long, few-flowered slender spikes of pink, purple and white flowers, each to 5 mm long, and linear pods to 5 cm long with upturned tip containing oblong dark brown seeds. (Fig. 60)

Vigna Savi

V. adenanthus (G.F.W.Mey.)Maréchal (*Phaseolus adenanthus*)

Rob 3187(S)

Localities recorded: An

Climbing herb with trifoliolate leaves, the leaflets ovate and to 14cm long, with few-flowered clusters of white flowers, each to 4cm long, and pods to 14cm long containing up to 15 reddish brown seeds.

V. angularis (Willd.)Ohwi & Ohashi (*Phaseolus angularis*)   'amberique gros, adzuki bean'

Bailey

No locality recorded

Bushy erect annual vegetable with trifoliolate leaves, short axillary sprays of bright yellow flowers and cylindrical straw coloured (or black or brown) pods containing up to 12 red, straw, brown or black seeds with a long white stalk-scar.

V. marina (Burm.)Merr.   'pois marron, shore bean'

Summerhayes, Fosberg (2)

Bru/War 39,182(S), Rob 2772(S)

Localities recorded: M,S,P,C,A,L,Ald,Cos

Indigenous strand plant, a scrambling perennial herb with trifoliolate leaves, the leaflets rounded and to 9.5cm long, yellow flowers to 1.5cm long on short axillary spikes and pods to 6cm long, slightly constricted between the seeds which are yellowish or reddish brown and up to 6 in number.

V. radiata (L.) Wilczek (*Phaseolus aureus*)   'amberique, green or golden gram, mung'

Bailey

No locality recorded

Semi-erect much branched hairy annual vegetable with trifoliolate leaves, axillary sprays of yellow flowers and long slender greyish or brown pods containing small spherical green, rarely yellow or black, seeds with a round white stalk-scar.

V. subterranea (L.)Verdc. (*Voandzeia subterranea*)   'bambara, ouange, bambara groundnut'

Bailey

No locality recorded

Cultivated annual herb with trifoliolate leaves, with elliptic leaflets to 8 cm long, rooting at the nodes, two pale yellow flowers, each to 8 mm long, in the leaf axils and rounded wrinkled pods to 2 cm long which develop underground and usually contain one round smooth hard edible seed.

V. unguiculata (L.)Walp. (*V. catjang, V. sinensis*)   'boeme, cow pea, black eye pea, yard long bean'

Summerhayes, Bailey, Fosberg (1),(2), Robertson (2)

Rob 2944(S)

Localities recorded: M(sr only),Co,Coe,Dar,Ald,Ast

Climbing annual or perennial herb with trifoliolate leaves, the leaflets ovate to 10 cm long, long stalked clusters of yellowish or pale violet flowers and long narrow cylindrical pods from 10 to 50 cm long and containing up to 20 beans or more.

ROSACEAE (58/1)

Eriobotrya Lindl.

E. japonica (Thunb.)Lindl.   'bibassier, loquat'

Bailey

No locality recorded

Small cultivated tree to 10 m with lanceolate leaves to 22 cm with prickly margins and pubescent beneath, fragrant white flowers to 2 cm wide in woolly clusters and round sweet edible yellow fruit to 3.5 cm long.

Fragaria L.

F. vesca L.   'fraisier, strawberry'

Bailey

Localities recorded: M(sr only)

Cultivated herb with trifoliolate leaves, the leaflets to 5cm long, white flowers to 2cm wide and pyramidal red juicy edible fruit in which the seeds are embedded. It also reproduces vegetatively by runners.

Prunus L.

P. persica (L.)Batsch   'pecher, peach, nectarine'

Bailey

No locality recorded

Cultivated small tree with lanceolate, serrated edged leaves to 15cm long, pink flowers to 3.5cm wide, usually before the leaves, and globose fleshy sweet edible fruit to 7cm wide containing one stone.

Rosa L.

R. spp.   'rosier, rose'

Bailey

Localities recorded: M(sr only)

Cultivated prickly shrubs with odd-pinnate leaves, the leaflets serrated edged and to 5 cm long, showy double or single flowers, variously coloured and some fragrant, and ovoid red fruit to 3 cm long containing many seeds. The nomenclature of these cultivated species is complicated.

Rubus L.

R. macrocarpus L.   'mure noire, ceylon blackberry'

Bailey

No locality recorded

Cultivated prickly scambling shrub with odd-pinnate leaves, white flowers and large purplish or red juicy edible fruits of many drupes on a core or receptacle.

R. niveus Thunb. (*R. lasiocarpus*)   'framboisier, wild raspberry, ceylon rasberry'

Bailey

No locality recorded

Cultivated straggling prickly shrub with white down on stems, odd-pinnate leaves, the leaflets ovate to rounded and white beneath and to 7 cm long, clusters of white or purplish flowers to 1 cm wide and hairy red juicy fruit to 1 cm wide composed of many small red drupes on a central core.

R. rosifolius Sm.   'framboisier, mauritius raspberry'

Summerhayes, Bailey

Jef 399,512(S), Sch 11784(K)

Localities recorded: M.S

Indigenous prickly shrub with stems with long white hairs, odd-pinnate leaves, the leaflets ovate, serrated edged to 5 cm long, solitary white flowers to 2.5 cm wide and elongated juicy red fruit to 2.5 cm long composed of hundreds of small drupes on a central core. (Fig. 61) (It is doubtful if all three *Rubus* species are present)

CHRYSOBALANACEAE (58/2)

Chrysobalanus L.

C. icaco L.   'coco plum, prune de france'

Bailey, Lionnet (1), Sauer, Wilson (1), Robertson (6)

Arc 154(K), Jef 406(S), Sch 11841,11735(K), Lam 419(K) Rob 2453,2569,2700(S)

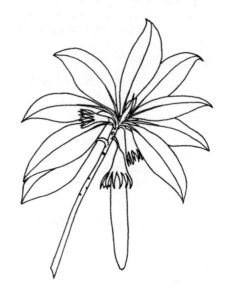

Fig. 61. Rubus rosifolius

Fig. 62. Bruguiera gymnorhiza

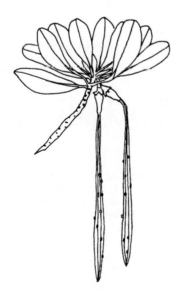

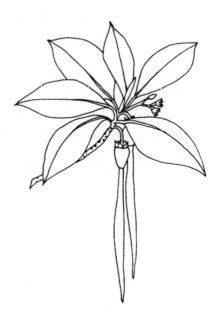

Fig. 63. Ceriops tagal

Fig. 64. Rhizophora mucronata

Localities recorded: M,P,F,N,C,An

Thicket-forming shrub with red stems with many lenticels, shiny ovate to rounded leaves to 10 cm long, white flowers to 6 mm long and globose red fruit to 3.5 cm wide containing one stone embedded in edible tasteless white flesh.

Parinari Aubl.

P. curatellifolia Planch. ex Benth.

Summerhayes

Localities recorded: M

Small tree with corky fissured bark part of the way up the trunk, ovate leaves with grey or rusty tomentum and prominent veins beneath, sprays of fragrant hairy pale mauve or white flowers, each to 4mm long, and ovoid edible yellow fruit drying to dark brown with corky patches.

HYDRANGEACEAE (59/6)

Hydrangea L.

H. macrophylla (Thunb.)Ser. ssp. **macrophylla** (*H. hortensis*)   'hortensia, french hydrangea'

Bailey

Localities recorded: M(sr only)

Cultivated shrub to 2 m with opposite ovate leaves to 20 cm long and dense rounded heads of white, pink or blue flowers, each to 2 cm wide.

BREXIACEAE (59/8)

Brexia Thouars

B. madagascariensis Thouars   'bois cateau'

Summerhayes, Bailey, Renvoize, Fosberg (2), Baumer

Dup 18/33(K), Ves 5541a,6119,6181,6193(K), Jef 556, 1143(S), 825(K), Pro 4088(S), Wil/Rob 2821(S)

Localities recorded: M,S,P,L,Ro,Pie,Ald

Indigenous shrub or small tree with the leaves variable, lanceolate and sinuate edged to 18 cm long or markedly toothed or entire obovate, greenish white flowers and cylindrical 5-ribbed fruit to 6 cm long, sometimes smooth and sometimes on long stalks. Taken as a charm in fishing pirogue and used medicinally in Seychelles.

CRASSULACEAE (60/1)

Bryophyllum Salisb.

B. pinnatum (Pers.)S.Kurz (*Kalanchoe pinnata*)   'sans sentiment, leaf of life'

Summerhayes, Bailey, Gwynne, Renvoize, Fosberg (1), Evans, Stoddart (1),(2),(3), Wilson (2),(4)

Lam 421(K), Pig sn(K), Bru/War 105(S), Rob 2350,2863,3038,3082,3175(S)

Localities recorded: M,P,D,C,A,Co,Bir,Den,Pla,Coe,Poi,Des,Dar, Jos,Mar,Alp,Far,Pie

Succulent herb with fleshy leaves, simple to odd-pinnate and to 25 cm long, erect spike of hanging reddish bell-shaped flowers, each to 4 cm long with a persistent calyx, and fruit rarely set, the plant reproducing vegetatively by means of bulbils from the leaf serrations, or by rooting from the nodes.

RHIZOPHORACEAE (65)

Bruguiera Lam.

B. gymnorhiza Lam.   'grand manglier, manglier latte, mangrove'

Summerhayes, Bailey, Lionnet (1), Renvoize, Sauer, Fosberg(2)

Jef 493(S)

Localities recorded: M,C,Ald,Cos,Ast

Indigenous shrub or tree, a mangrove, which may develop a buttressed trunk, and with knee-like aerial roots arising from the mud, with opposite elliptical leathery leaves to 15 cm long at the ends of the branches, solitary axillary flowers to 2 cm long, hanging down and with a many-lobed reddish calyx and whitish petals followed by the cigar-shaped hypocotyl to 25 cm long. (Fig. 62)

Cassipourea Aubl.

C. gummiflua Tul. (*C. paradoxa*)

Summerhayes, Procter (1)

No locality recorded

Tree or shrub with variable ovate leaves to 16 cm long, small hairy flowers to 8 mm long in clusters in the leaf axils and ovoid capsules. Probably an introduced species.

C. thomassettii (Hemsley)Alston

Renvoize, Fosberg (2)

Localities recorded: Ald

Endemic small tree with elliptic leaves to 11 cm long, greyish above and brown below, clusters of small hairy flowers in the leaf axils, each to 4 mm long, and the fruit unknown.

Ceriops Arn.

C. tagal (Perr.)C.B.Rob. (*C. candolleana*), 'manglier jaune, manglier giroflier, mangrove'

Summerhayes, Renvoize, Fosberg (2), Wilson (2)

Arc 131(K), Wil/Rob 2788(S)

Localities recorded: M,C,Ald,Ass,Cos

Indigenous shrub or small tree, a mangrove, with opposite elliptical leaves to
8 cm long, axillary clusters of small white flowers, each to 6 mm long, and a
strongly ribbed hypocotyl to 25 cm long from a conical fruit to 1.5cm long.
The trunk has a pyramidal base and aerial roots project finger-like from the
mud. (Fig. 63)

Rhizophora L.

R. mucronata Lam.   'manglier gros poumon, manglier hauban, mangrove'

Summerhayes, Bailey, Renvoize, Sauer, Lionnet (1), Fosberg (2), Stoddart
(1),(2)

Jef 495(S), Pro 4281(S), Rob 2359(S)

Localities recorded: M,S,P,C,Poi,Jos,Far,Ald,Cos,Ast

Indigenous tree to 25 m, a mangrove, with aerial bowed prop roots arising
from quite high up the trunk, leathery elliptical opposite leaves to 17 cm long,
axillary fleshy yellow flowers to 1.5 cm long and a slender hypocotyl to 40 cm
long, slightly swollen about the centre, from a 4 cm long pear-shaped fruit.
(Fig. 64)

COMBRETACEAE (66)

Lumnitzera Willd.

L. racemosa Willd.   'manglier à petite feuilles'

Summerhayes, Renvoize, Fosberg (2)

Jef 789(S), Rob 2289(S), Tod 37(S)

Localities recorded: M,P,F,Ald,Ast

Indigenous small tree with rough bark and fleshy sessile oblong leaves to 7
cm long, spirally arranged, with axillary spikes of sessile white or cream
flowers, each to 1 cm long, with a fleshy compressed tubular calyx with
bracteoles and compressed ovate fruit to 1.2 cm long. A mangrove. (Fig. 65)

Quisqualis L.

Q. indica L.   'liane vermifuge, orientale, santonine, rangoon creeper'

Summerhayes, Bailey, Wilson (1)

Jef 610(S), Fos 52132(S)

Localities recorded: M,P,Co

Semi-climbing shrub with opposite leaves to 12 cm long, the petioles
becoming thorns, and drooping heads of long tubed pink or red sweetly
scented flowers, the tube to 7 cm long and the limb to 4.5 cm wide. (Fig. 66)

Fig. 65. Lumnitzera racemosa

Fig. 66. Quisqualis indica

Fig. 67. Callistemon speciosus

Fig. 68. Eucalyptus camaldulensis

Terminalia L.

T. arjuna (Roxb.)Wight & Arn.   'arjun'

Swabey

Localities recorded: M

Large timber tree with buttressed trunk and drooping branches, sprays of small white flowers and fruit to 5 cm long with 5 to 7 wings down the fruit each to 1cm wide.

T. bellirica (Gaertn.)Roxb. (not *T. balerica*)

Swabey

Localities recorded: M

Large timber tree to 60 m, mature trees with buttressed trunks, with long-petioled elliptical leaves to 20 cm long at the ends of the branchlets, clusters of greenish white flowers smelling of honey and grey velvety tomentose globose to ovoid fruits to 3 cm long with 5 ridges, the stone embedded in yellow flesh.

T. boivinii Tul.   'bois faune'

Renvoize, Fosberg (2)

Localities recorded: Ald,Ass,Ast

Small tree to 3 m with obovate leaves to 6 cm long, 4 cm long spikes of yellowish flowers clustered on spur shoots, and ellipsoid fruit to 12 mm long.

T. catappa L.   'badamier, indian almond'

Summerhayes, Sumner Gibson, Bailey, Piggott, Gwynne, Lionnet (1), Renvoize, Fosberg (1),(2), Stoddart (1), (2),(3), Baumer, Wilson (1),(2), Robertson (3),(4),(6)

Jef 369,381(S), Pro 4022(S), Bru/War 71(S), Rob 3138(S)

Localities recorded: M,P,F,C,Fe,An,?Co,Bir,Den,Coe,Rem,Des,Dar,Jos, Poi,Pro,Far,Ald,Ass,Cos,Ast

Indigenous spreading tree to 15 m with young branches horizontal, large obovate leaves to 40 cm long, turning red before falling, pendulous spikes to 15 cm long of small yellowish flowers and ellipsoid compressed fruit to 6 cm long, dark maroon when ripe and containing one nut. Used medicinally in Seychelles.

MYRTACEAE (67/1)

Callistemon R.Br.

C. citrinus (Curtis)Stapf   'bottle brush'

Lionnet (2)

Localities recorded: M(BG)

Cultivated shrub to 8 m with lanceolate leaves to 7 cm long and 10 cm long whorls near the ends of branches of crimson flowers with long stamens and fruit to 7 mm wide.

C. speciosus (Sims)DC.   'bottle brush'

Robertson (7)

Jef 740(S)

Localities recorded: M,Dar

Cultivated shrub, usually less than 8 m tall, with narrowly lanceolate leaves to 17 cm long, 30 cm long whorls of red, many-stamened flowers near the ends of the branches with the small cup-like fruit to 5 mm wide. (Fig. 67) (There is likely to be one or other species on Mahe, but not both.)

Eucalyptus L'Hérit.

E. alba Reinw. ex Reinw. ex Blume   'eucalyptus, timor white gum'

Bailey

No locality recorded

Tall tree with smooth white bark and slender drooping branches, flowers in 7-flowered heads, the bud cap hemispherical, and fruit hemispherical to 1 cm wide.

E. camaldulensis Dehn.   'eucalyptus, murray red gum'

Bailey

Jef 1161(S), Fos 52150,52137(S), Pro 3951(S), Bru/War 88(S)

Localities recorded: M,?A,Co

Tall tree with smooth ash coloured bark, alternate lanceolate leaves, heads with up to 10 flowers, the bud cap long conical, and hemispherical fruit to 9 mm wide. (Fig. 68)

E. citriodora Hook.   'eucalyptus, lemon scented gum'

Sumner Gibson, Bailey

Localities recorded: M

Tall slender tree to 20 m with reddish grey peeling bark on a spotted trunk, pendulous branches of dense lemon-scented foliage, clusters of up to 5 cream flowers, the bud cap conical, and urn-shaped fruit to 7 mm wide.

E. globulus Labill.   'eucalyptus, tasmanian blue gum'

Bailey

No locality recorded

Tall tree with smooth bark, alternate sickle-shaped leaves and 4-angled young shoots, solitary flowers with a warty bud cap and sessile ridged top-shaped fruit to 2.8 cm wide.

E. maculata Hook.f.   'spotted gum'

Dup 1931(K)

Localities recorded: ?M

Tree to 50 m with smooth white bark which is shed in large patches, alternate lanceolate glossy leaves, clusters of up to 5 flowers, the bud cap hemispherical, and urn-shaped fruit to 1.5 cm wide.

E. robusta Sm.   'eucalyptus, iron bark, swamp mahogany'

Bailey, Sumner Gibson

Pro 3971,3981(K)

Localities recorded: M

Tall spreading tree with rough brown fibrous bark, the young twigs reddish, alternate lanceolate leaves with long points, cream flowers to 2 cm long and up to 10 per cluster, the bud cap short and conical, and urn-shaped fruit to 1.5 cm long.

E. staigeriana F.Muell. ex F.M.Bailey   'eucalyptus'

Bailey, Sumner Gibson

Jef 680(S), Pro 3953(S), 4146(K)

Localities recorded: M

Tree to 8 m with dark rugose bark, narrow lanceolate leaves, clusters of up to 8 flowers, each to 8 mm long, the bud cap conical, and globose fruit to 5 mm wide. (Fig. 69)

E. tereticornis Sm.   'eucalyptus locale, forest red gum'

Bailey, Sumner Gibson

Localities recorded: ?M

Tall tree with irregularly blotched bark, alternate lanceolate glossy leaves, clusters of up to 12 flowers, each to 1.2 cm long, the bud cap long-conical, and hemispherical fruit to 6 mm wide.

Eugenia Mich. ex L. (*Syzygium*)

E. aquea L.   'watery rose apple'

Summerhayes, Wilson (2)

Fos 52114(S)

Localities recorded: M,C,Co

Large cultivated tree with glabrous opposite elliptic leaves to 20 cm long, whitish many-stamened flowers and edible fruit. (The specimens and records quoted may be *E. malaccensis*.)

E. caryophyllus (Spreng.)Bull. & Harr. (*E. caryophyllata*)   'girofleur, clove'

Bailey, Lionnet(1)

Fig. 69. Eucalyptus staigeriana   Fig. 70. Psidium guajava

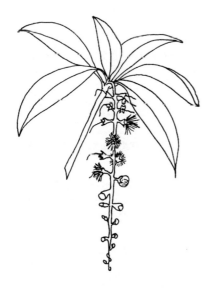

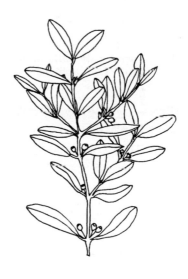

Fig. 71. Barringtonia racemosa   Fig. 72. Memecylon eleagni

Localities recorded: M,S,P

Cultivated and escaping columnar tree with erect branches to 20 m, opposite shining aromatic leaves, paler below, ovate to 10 cm long, the young leaves pink, flowers in clusters with the flower-stem and calyx turning red, the petals falling very early, and many greyish yellow stamens, and ovoid dark purple one-seeded fruit to 2.5 cm long called 'mother of cloves'. The commercial spice is the dried unopened flower bud and stem.

E. cuminii (L.)Druce   'jamblong, jambolan, java plum'

Bailey

Bru/War 48(S)

Localities recorded: A

Cultivated and escaping spreading evergreen tree with opposite ovate leaves to 15 cm long, many-stamened flowers to 8 mm long and edible acid ovoid purple fruit to 3 cm long.

E. elliptica Lam. var. **levinervis** Fosberg

Fosberg (2)

Localities recorded: Ald

Endemic glabrous shrub with opposite elliptic leaves to 10 cm long and globose purple black fruit to 1 cm wide.

E. grandis Wight

Dup 11/33(K), Jef 842(S)

Localities recorded: M

Tall tree with scaly bark, opposite ovate leathery leaves to 25 cm long and dense heads of white flowers, each to 2 cm long.

E. jambos L.   'jambrosa, rose apple'

Summerhayes, Bailey, Lionnet (1), Wilson (1)

Jef 403,841(S), Gwy/Woo 1076(K), Sch 11789(K), Pro 3920(S), Rob 2421(S)

Localities recorded: M,S,P

Cultivated and escaping tree with lanceolate leaves to 20 cm long, dark green above, flowers to 4 cm long with many stamens and pale yellow petals, and round fleshy pale yellow rose-scented edible fruit to 3 cm wide.

E. javanica Lam.   'jamalac rouge, jamalac blanc, java apple'

Bailey, Robertson (4)

Sch 11766(K)

Localities recorded: M(sr only),P,F

Cultivated small spreading tree with ovate leaves to 20 cm long, greenish white flowers with yellow stamens to 2.5 long and clusters of waxy rose pink or greenish white pear-shaped edible fruit to 4 cm wide.

E. malaccensis L. 'pomme gouverment, malay apple, pomerac'

Bailey, Lionnet (1),(2), Wilson (1),(2), Robertson (4),(5)

Jef 743(S), Bru/War 101(S)

Localities recorded: M(BG),P,F,N,C,A

Cultivated and escaping large tree with oblong leaves to 40 cm long, red flowers with brilliant magenta stamens which make a carpet beneath the tree, and bright red ovoid or pear-shaped fruit to 8 cm long with edible white flesh and one seed.

E. uniflora L. (*E. michelii*) 'roussaille, pitanga or surinam cherry'

Summerhayes, Bailey

Pro 4278,4478(K), Rob 2598(S)

Localities recorded: M,N

Cultivated and escaping shrub with ovate pointed shiny leaves to 10 cm long, red when young, small white flowers and flat round ribbed fruit to 4 cm wide which are edible and made into jam. Used as a hedge plant.

E. wrightii Baker (*E. sechellarum*) 'bois de pomme'

Summerhayes, Bailey, Procter (1), Baumer, Wilson (1)

Ves 5484,5430,5517,5541,5543,5773,6363(K), Jef 554, 627(K), Pro 3989, 4019(S),4573(K), Rob 2268,2774(S)

Localities recorded: M,S,P,Fe,C

Endemic tree or untidy scrambling shrub with leaves variable but usually broadly ovate and leathery to 20 cm long, white many-stamened flowers to 2 cm long and globose black fruit to 2 cm wide with red flesh and one seed. Used medicinally in Seychelles.

Melaleuca L.

M. quinquenervia (Cav.)S.T.Blake 'cajaput, cayapouti, punk tree'

Bailey (cultivated plants known as *M. leucadendron* are this species)

No locality recorded

Cultivated tree to 10 m with loose white spongy scaling bark, pendulous branchlets, lanceolate leaves to 8 cm long near the ends of the branches, dense spikes of creamy white flowers like bottle brushes and sessile cup-shaped fruit to 5 mm wide often fused together on the stem. Cajeput oil can be distilled from the leaves.

Pimenta Lindl.

P. dioica (L.)Merr. (*P. officinalis*)   'quatre-epice, allspice, pimento'

Bailey

Ves Dec/36(K), Pro 4363(K)

Localities recorded: M(BG)

Small tree to 12 m with rounded crown, silvery peeling bark, elliptic leaves to 15 cm long, paler below, clusters of creamy white flowers, each to 6 mm wide, and small globose fruit to 6 mm wide, ripening dark purple. The spice is the dried unripe fruit.

Psidium L.

P. guajava L.   'goyavier, guava'

Summerhayes, Bailey, Baumer, Wilson (1),(2), Robertson (5)

Localities recorded: M(sr only),P,N,C

Cultivated shrub or small tree to 10 m with reddish brown bark peeling off in thin flakes, oblong leaves to 15 cm, thinly pubescent below, the young shoots narrowly winged, many-stamened white flowers to 4 cm wide and globose fruit to 10 cm wide with yellow skin, pink or white edible flesh and many hard irregularly shaped seeds. Used medicinally in Seychelles. (Fig. 70)

P. littorale Raddi (*P. cattleianum*)   'goyavier de chine, goyavier marron, chinese or strawberry guava'

Summerhayes, Bailey, Wilson (1), Robertson (4),(5)

Jef 675(S), Bru/War 100(S)

Localities recorded: M,P,F,N,A

Small tree to 8 m with reddish peeling bark, thick shiny obovate leaves to 10 cm, white many-stamened flowers to 2.5 cm wide and globose red or yellow juicy edible fruit to 3.5 cm wide.

LECYTHIDACEAE (67/2)

Barringtonia Forst.

B. acutangula Gaertn.

Summerhayes

No locality recorded

Indigenous tall tree with oblanceolate leaves to 18 cm long, crowded at the ends of the branches, flowers to 2 cm wide on hanging racemes to 25 cm long and oblong sharply angled fruit to 3.5 cm long.

B. asiatica (L.)Kurz   'bonnet carre bord mer'

Summerhayes, Bailey, Gwynne, Fosberg, Lionnet (1), Renvoize, Stoddart (1),(2),(3), Wilson (2),(4), Robertson (4)

Jef 735(S), Bru/War 68(S), Rob 2618,2834,3103(S)

Localities recorded: M,P,F,N,C,A,Co,Den,Coe,Des,Dar,Jos,Mar,Alp,Far

Indigenous large spreading tree with sessile shiny leaves to 40 cm long, flowers opening at night with 2 large sepals, 4 white petals and many long stamens, half pink, half white, to 12 cm long, and large squarish fruit to 12 cm wide with one seed embedded in fibrous tissue and which can float in the sea.

B. racemosa (L.)Spreng.   'bonnet carre de riviere, bois marai grand feuille'

Summerhayes, Bailey, Lionnet(1)

Rob 2547(S), Jef 788(K)

Localities recorded: M,S,P

Indigenous tree with ovate leaves to 20 cm with pendant spikes of pinkish flowers each to 5 cm wide, to 40 cm long, and fibrous angular fruit. (Fig. 71)

Couroupita Aubl.

C. guianensis Aubl.   'canon ball tree'

Bailey, Lionnet (2)

Localities recorded: M(BG)

Cultivated tall tree to 20 m with oblong leaves to 25 cm long, the veins beneath pubescent, large flowers in clusters up the trunk, each to 10 cm wide with white petals and a wide staminal column like a bent spoon, and large brown globose fruit to 15 cm wide. (Little boys in the Botanic Garden mistakenly tell visitors that this is a 'fly trap')

Gustavia L.

G. angusta L.   'stink wood'

Jef 1163(S)

Localities recorded: M(BG)

Tree with large pointed ovate sparsely toothed leaves to 40 cm long, scented pink flowers to 20 cm wide with dense ring of stamens with pink filaments.

MELASTOMATACEAE (68)

Melastoma Burm. ex L.

M. malabathricum L.   'quatouc, watouc, indian rhododendron'

Summerhayes, Bailey

Arc 145(K), Jef 567,826(K),666(S), Sch 11656,11781(K), Fos 52027(S), Rob 2544(S)

Localities recorded: M,S

Indigenous woody herb to 2 m with hairy leaves to 12 cm long, 3–5-nerved from the base, white flowers to 4.5 cm wide fading to pink and fruit to 6mm wide with edible reddish pulp.

Memecylon L.

M. eleagni Blume   'bois calou'

Summerhayes, Bailey, Procter (1)

Ves 5428,5544,5503,5646,6118(K),Arc 165(K),Jef 642,730(S),Sch 11826,11689(K), Fos 51977(S), Pro 3931, 3954(S), Ber 14632(K), Rob 2272(S)

Localities recorded: M,S,P,D,C,Fe

Endemic shrub with opposite ovate shiny leaves to 6 cm long, held erect, small white axillary flowers to 5 mm wide and shortly stalked globose fruit to 8 mm long, ripening dark blue or black. (Fig. 72)

M. floribundum Blume (*M.caeruleum*)

Dup 2/31(K), Jef 609(S)

Localities recorded: M

Shrub to 3 m with glossy sessile opposite ovate leaves to 15 cm long, clusters of small flowers, each to 6 mm long, in the leaf axils, the sepals lilac and the petals deep blue, and globose fruit, pink at first, then black, to 1.5 cm long.

M. umbellatum Burm.

Lionnet (2)

Localities recorded: M(BG)

Cultivated small tree to 8 m with leathery shiny oblong leaves to 14 cm long, clusters of small deep blue or purplish flowers on the branches and globose dark purple fleshy fruit to 8 mm wide.

LYTHRACEAE (69/1)

Lagerstroemia L.

L. indica L.   'goyavier fleurs, indian lilac, crape myrtle'

Bailey

Localities recorded: M(sr only)

Cultivated shrub to 7 m with dark green sessile subopposite ovate leaves to 6 cm long, the young shoots 4-angled, and terminal sprays of frilly-petalled rosy pink or mauve flowers, each to 2.5 cm long.

L. speciosa (L.)Pers. (*L. flos-reginae*)   'pride of india, queen's flower'

Bailey

No locality recorded

Cultivated tall, much branched tree to 30 m with subopposite oblong leaves to 30 cm long, erect panicles of showy mauve or pink flowers, each to 3.5 cm long and flowering before the leaves, and globose capsules to 1.5 cm wide with a woody star shaped calyx.

Lawsonia L.

L. inermis L. (*L.alba*)   'henne, reseda, henna'

Bailey, Robertson (7)

Jef 1196(S)

Localities recorded: M,Dar

Shrub with sessile opposite elliptic leaves to 2.5 cm long, terminal sprays of small creamy white or rose pink fragrant flowers, each to 6 mm wide, and globose capsules to 8 mm wide. The soaked leaves yield the dye. (Fig. 73)

Pemphis J.R.& G.Forst.

P. acidula Forst.   'bois d'amande'

Summerhayes, Gwynne, Piggott, Renvoize, Fosberg (2), Stoddart (1),(2)

Jef 1207(S), Rob 2385(S)

Localities recorded: Bir,Den,Rem,Jos,Poi,Pie,Far,Ald, Ass,Cos,Ast

Indigenous shrub or small tree of beach crest, much branched with ovate lanceolate grey green leathery leaves to 2 cm long crowded towards the ends of the branches, solitary white or cream flowers to 8mm wide and globose capsules sunk in the calyx, each to 8 mm long and containing reddish brown winged seeds. Possibly confused in some localities with *Suriana maritima* as Stoddart (3) says it is not on Bird or Denis. (Fig. 74)

SONNERATIACEAE (69/3)

Sonneratia L.

S. alba Sm. (*S. caseolaris*)   'manglier fleurs'

Summerhayes, Renvoize, Fosberg (2)

Localities recorded: M,Ald,Cos

Indigenous small tree in mangrove swamps with reddish trunk, stout finger-like aerial roots coming out of the mud, opposite obovate leaves to 10cm long, scented white flowers to 2cm wide, solitary or in 3's at the shoot apex, and green flattened-globose fruit to 3cm wide with persistent style base and variable calyx lobes, containing many seeds. (Fig. 75)

PUNICACEAE (69/4)

Punica L.

P. granatum L.   'grenadier, pomegranate'

Bailey, Wilson (2), Robertson (7)

Localities recorded: M(sr only),C,Dar

Cultivated shrub or small tree to 8 m with opposite shiny oblong leaves to 7 cm long, fleshy orange red flowers to 3 cm long and large globose fruit to 12 cm wide with yellow leathery skin and numerous seeds in pink translucent edible flesh. (Fig. 76)

ONAGRACEAE (70/1)

Ludwigia L.

L. erecta (L.)Hara

Summerhayes (in part; G, HPT 21, Horne 370, Perville)

Jef 630(S),Pro 4137(S),Bru/War 30(S),Rob 2556,2648(S)

Localities recorded: M,S,D,F,N,A,Fe

Glabrous herb with reddish stems, lanceolate leaves to 15 cm long, axillary yellow flowers to 8 mm wide and elongated 4-ribbed capsules to 2 cm long with the 4 sepals persistent at the apex.

L. octovalvis (Jacq.)Raven incl. ssp. *sessiliflora* (Mich.)Raven and ssp. *octovalvis* (*Jussiaea suffruticosa*)   'herbe la mare'

Summerhayes,Bailey,Wilson (1),(2),Rob (5)

Arc 159(K),Jef 376(S),Sch 11728(K),Lam 418(K),Fos 52116(S),Pro 4134(S), 4382(K),Rob 2400(S)

Localities recorded: M,S,P,N,C,L,Co

Softly hairy herb of wet places with ribbed stems, opposite ovate leaves to 10 cm long, yellow flowers to 2 cm wide and 4-ribbed seed capsule.

L. jussiaeoides Desr.

Summerhayes (in part; G.Horne 456, Perville)

Localities recorded: M

Indigenous herb similar to *L. erecta* but with puberulous young shoots and often larger flowers. (Pro 4137 and Bru/War 30 may belong here).

TURNERACEAE (73)

Turnera L.

T. ulmifolia L.   'la coquette, yellow alder'

Summerhayes,Bailey,Gwynne,Lionnet(1),Renvoize,Fosberg(2),Stoddart(1), (2),(3),Baumer,Wilson(1),(2),(4),Robertson(6)

Ves 5456(K),Arc 152(K),Jef 528(S),Pig sn,sn(K),Fos 52035(S),Sch 11652(K),Pro 4341(K),Rob 2325,2355,2394,2568,2654,2854,3046,3068(S)

Localities recorded: M,S,P,F,N,C,An,L,Co,Bir,Den,Pla,Coe,Rem,Des,Dar, Jos,Poi,Mar,Alp,Pie,Far,Ald,Cos,Ast

Herb, sometimes woody and shrubby, with pubescent toothed ovate lanceolate leaves to 15 cm long, large yellow flowers to 5 cm wide at the top of

Fig. 73. Lawsonia inermis

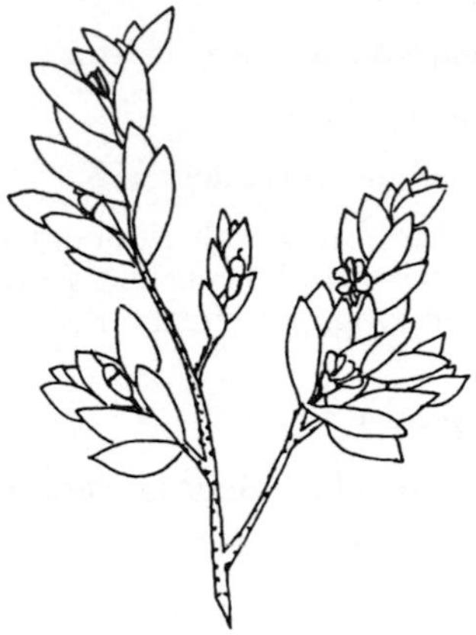

Fig. 74. Pemphis acidula

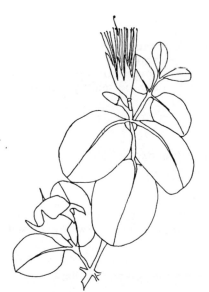

Fig. 75. Sonneratia alba

Fig. 76. Punica granatum

the plant, closing by midday, and conical capsules containing many irregular yellow brown seeds. Used medicinally in Seychelles. (Fig. 77).

PASSIFLORACEAE (74/1)

Adenia Forssk.

A. gummifera (Harv.)Harms   'la liane maria'

Summerhayes

Localities recorded: M,S

Indigenous woody climber with simple or lobed leaves to 11 cm long, tendrils, ♂ flowers to 1.5 cm wide, ♀ flowers to 1 cm wide and up to 4 ovoid fruit, each to 3 cm long, per cluster. Formerly the sap was used as a drink.

Passiflora L.

P. caerulea L.   'fleur de la passion, blue passion flower, passion vine'

Bailey

Localities recorded: M(sr only)

Cultivated climbing vine with 3–5-lobed leaves, bluish purple large showy flowers to 10 cm wide and with 4 rings of filaments, and ovoid yellow fruit to 6 cm long.

P. edulis Sims (incl. var. *flavicarpa* Degener)   'passion fruit, grenadilla'

Bailey

Localities recorded: M(sr only), P(sr only)

Cultivated vine with 3-lobed leaves to 15 cm long, axillary tendrils, flowers to 10 cm wide with white petals, a row of wavy purple and white corona filaments and 5 stamens joined to the 3-lobed stigma, opening in the afternoon, and globose edible fruit with black seeds in a juicy pulp, normally purplish but yellow in the var. *flavicarpa*. (Fig. 78)

P. foetida L.   'poc-poc, wild passion fruit'

Bailey, Fosberg (2), Stoddart (1), Wilson (1), Robertson (4),(6)

Jef 377(S), Fos 52151,52083(S), Rob 2469(S),2576(K)

Localities recorded: M,P,F,N,An,Co,Far,Ass

Herbaceous climber with 3-lobed leaves to 12 cm long, flowers to 5 cm wide, white with lower part of corona purple, and orange fruit to 2 cm wide enclosed in a cage of expanded feathery bracts.

P. laurifolia L.   'grenadille, water lemon, belle apple, pomme de liane'

Bailey

No locality recorded

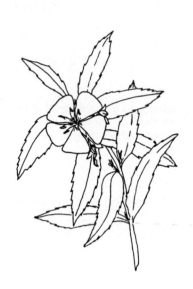

Fig. 77. Turnera ulmifolia

Fig. 78. Passiflora edulis

Fig. 79. Cucumis anguria

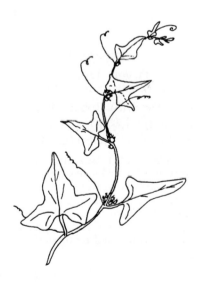

Fig. 80. Mukia maderaspatana

Glabrous woody climber with entire leathery leaves to 14 cm long, large fragrant flowers to 10 cm wide, white dotted with rosy purple and purple banded corona, and ovoid orange yellow fruit to 8 cm long with edible pulp around the seeds.

P. quadrangularis L. 'grenadine'

Summerhayes, Bailey

Localities recorded: M

Cultivated strong glabrous climber with 4-angled and winged stem, axillary tendrils, ovate leaves to 25 cm long, white flowers to 12 cm wide, tinged red and purple, and ovate yellow fruit to 30 cm long with edible white flesh which is sometimes cooked as a vegetable when unripe.

P. suberosa L. 'lipeca'

Summerhayes, Gwynne, Renvoize, Fosberg (2), Stoddart (1),(2),(3), Wilson (1),(2),(3),(4), Robertson (3),(6)

Jef 440(S),Fos 52169(S),Pro 4067(S),Fra 437(S),Rob 2484, 2550,2831,3042,3090(S)

Localities recorded: M,P,N,C,An,L,Co,Bir,Den,Pla,Coe,Rem,Des,Dar,Jos,Poi, Mar,Def, Alp,Far,Ald,Ass,Cos,Ast

Vigorous herbaceous climber with leaves variable from simple ovate to 3-lobed to 9 cm long, greenish flowers to 1.5 cm wide and dark blue or black ovoid fruit to 1.3 cm wide.

CARICACEAE (74/4)

Carica L.

C. papaya L. 'papayer, pawpaw, melon tree'

Bailey, Piggott, Lionnet (1), Fosberg (1),(2), Stoddart (1),(2),(3), Baumer, Wilson (2),(3),(4), Robertson (2),(3),(4),(5)

Bru/War 87(S), Rob 2824(S)

Localities recorded: M(sr only),F,N,C,A,Co,Bir,Den,Coe,Rem,Des,Dar, Poi,Mar,Def, Alp,Pro,Far,Ald

Cultivated and escaping stout sparsely branched thick-trunked tree with prominent trunk scars, palmately lobed and toothed glabrous leaves to 70 cm wide, various arrangements of small ♂ and large ♀ flowers, creamy white, and ovoid orange fleshy edible fruit containing many black mucilage-covered seeds, to 30 cm or more long. The unripe fruit can be cooked as a vegetable and the meat tenderizer papain is obtained from the milky sap of the cut unripe fruit. Used medicinally in Seychelles.

CUCURBITACEAE (75)

Citrullus Schrad.

C. lanatus (Thunb.)Matsum. & Nakai 'melon d'eau, water melon'

Bailey

No locality recorded

Slender hairy sprawling annual, cultivated, with 2–3-fid tendrils, pinnately lobed leaves, solitary ♂ or ♀ yellow flowers to 3 cm wide and large globose or oblong fruit to 60 cm long with hard green and cream mottled rind and usually red juicy edible flesh with many flat ovoid pointed black seeds.

Cucumis L.

C. anguria L.   'concombre, concombre marron, west indian gherkin'

Bailey, Fosberg (2)

Jef 1185(K)

Localities recorded: M(sr only),F,Ast

Slender trailing annual hairy herb with simple tendrils, palmately lobed leaves, yellow flowers to 1 cm wide and oval fruit to 5 cm long, covered with sharp hairs on warts, and ripening from pale green to ivory with small white seeds. (Fig. 79)

C. melo L.   'melon de france, canteloupe melon, vatangue'

Summerhayes, Bailey, Gwynne, Fosberg (2), Robertson (4)

Localities recorded: F,R,Coe,Ass,Cos

Cultivated trailing softly hairy annual with simple tendrils, shallowly lobed leaves, axillary clusters of yellow flowers to 3 cm wide and variable fruit, globose or oblong, rind smooth or reticulate, edible flesh yellow, pink or green, but all with many flat smooth elliptic seeds, whitish, to 15 mm long.

C. metuliferus E.Mey. ex Schrad.   'concombre, african horned cucumber'

Bailey

No locality recorded

Trailing hairy annual with heart-shaped 3-lobed leaves with yellow flowers to 3 cm wide and oblong spiny red fruit to 12 cm long.

C. sativus L.   'concombre, cornichon, cucumber'

Bailey, Robertson (4)

Localities recorded: M(sr only),F

Cultivated trailing hairy annual with simple tendrils, palmately angled leaves, yellow ♂ and ♀ flowers to 4 cm wide in axillary clusters and oblong or elongated fruit with a few spinous tubercles when young. The fruit is usually eaten as a salad vegetable when unripe, but ripens to yellow and then contains flat elliptic white seeds to 1 cm long.

Cucurbita L.

C. maxima Duch. ex Lam.   'squash'

Stoddart (3)

Localities recorded: Bir,Den (srs only)

Cultivated herbaceous vine similar to *C. moschata* but with rounded bristly stems and the fruit stalk corky, not enlarged at the fruit.

C. moschata (Duch. ex Lam.)Duch. ex Poir   'giraumon, patisson, pumpkin'

Bailey, Fosberg (1),(2), Stoddart (1),(3), Wilson (4), Robertson (3),(4)

Sto 7201(K), Sto/Poo 1323(K)

Localities recorded: M(sr only),F,Co,Bir,Dar,Poi,Mar,Far,Ald,Cos

Cultivated annual softly pilose vine, with angular hard stems, branched tendrils, rounded leaves to 30 cm long, solitary orange ♂ and ♀ flowers to 12 cm long and large elongated or flattened globose fruit, green and white mottled but ripening to buff and containing orange flesh, cooked as a vegetable, and white plump ovate seeds with an edge. The fruit stalk is enlarged at the fruit.

C. pepo L.   'marrow, courgette'

Gwynne, Stoddart (1)

Localities recorded: M(sr only), Far

Cultivated annual herb with short stout stem at first, then trailing, sharp bristly stem, large heart-shaped lobed leaves, to 50 cm long, solitary orange ♂ and ♀ flowers to 10 cm long, and oblong fruit, mottled green and white and ripening yellow with yellow flesh and smooth fat white ovate seeds to 1 cm long. Very young fruit are eaten cooked as courgettes, and the mature fruit is eaten cooked as marrow.

Lagenaria Ser.

L. siceraria (Molina)Standl.   'calebasse, bottle gourd'

Bailey, Fosberg (2)

Localities recorded: M(sr only),Ald

Annual trailing musky scented cultivated herb with bifid tendrils, hairy heart-shaped leaves, solitary axillary white flowers to 10 cm wide and club-shaped fruit with a hard rind to 1 m long containing compressed ridged seeds to 2 cm long. The shell can be used as a utensil and non-bitter varieties are eaten.

Luffa Mill.

L. acutangula (L.)Roxb.   'pipengaye a cotes, angled loofah'

Bailey

No locality recorded

Stout climbing annual, foetid if bruised, with 3-fid or more tendrils, lobed leaves, ♂ flowers in clusters and ♀ flowers solitary in the same leaf axil, yellow to 5 cm wide and club-shaped fruit to 50 cm long with 10 ribs and containing black pitted seeds to 1.4 cm long. Immature fruits used as a vegetable.

L. cylindrica (L.)M.J.Roem.   'pipengaye uni, smooth loofah, vegetable sponge'
Bailey

No locality recorded

Vigorous annual climber with 3-fid or more tendrils, large 7-lobed leaves, clusters of ♂ flowers and a single ♀ flower in each leaf axil, yellow to 10 cm wide, and cylindrical fruit, slightly ribbed or furrowed, to 60 cm long with smooth black seeds. Immature fruits can be eaten and the vascular network of the mature fruit is used as a sponge.

Momordica L.

M. charantia L.   'margoze, bitter gourd'

Summerhayes, Bailey, Fosberg (1),(2), Robertson (2), (3)

Localities recorded: M,Co,Coe,Poi,Ald,Ass

Slender annual climber with bifid or simple tendrils, leaves palmately 9-lobed, solitary axillary yellow flowers to 3 cm wide and hanging ribbed and tubercled oblong fruit to 25 cm long with numerous seeds which are brown with a red aril. The unripe fruits are eaten as a vegetable and in curries.

Mukia Arn.

M. maderaspatana (L.)M.J.Roem. (*Melothria maderaspatana*)   'cornichon sauvage'

Summerhayes, Gwynne, Renvoize, Stoddart (1),(3)

Localities recorded: Den,Far

Hairy perennial trailer with triangular leaves, sometimes 3-lobed, to 4 cm long, simple tendrils, clusters of ♂ and ♀ flowers, yellow, and red globose fruit to 1.1 cm wide. (Fig. 80)

Peponium Engl.

P. sublitorale C.Jeffrey & J.S.Page

Renvoize, Fosberg (2)

?Horne 460(K), ?Bru/War 66(S)

Localities recorded: ?M,?A,Ald

Endemic trailing plant with shallowly 5-lobed leaves to 15 cm wide, simple tendrils, clustered ♂ and solitary ♀ flowers, yellow, and ellipsoid pubescent fruit to 6 cm long containing ovate flattened seeds to 6 mm long. (The specimens from Mahe and Aride may be a different species)

Sechium P.Br.

S. edule (Jacq.)Swartz   'choucoute, choyote, christophine'
Bailey

Localities recorded: M(sr only)

Robust slightly hairy cultivated vine with large branched tendrils, wide ovate leaves, small green or cream flowers the ♂s in clusters, the ♀s solitary in the same leaf axil, and fleshy pear shaped green fruit to 20 cm long with one large flat seed. The fruit is cooked as a vegetable.

Trichosanthes L.

T. cucumerina L. (*T.anguina*) 'patole, snake gourd'

Bailey, Fosberg (2), Robertson (2),(4)

Localities recorded: M(sr only),F,Coe,Ald

Cultivated climbing annual herb with branched tendrils, angular 5-lobed leaves, ♂ flowers clustered, ♀ flowers solitary, white to 5 cm wide and with long white hairs on the petals, and long slender fruit to 75 cm or more, white or green and white with thick brown seeds embedded in woolly pulp. The flesh is green and eaten cooked or raw. (Fig. 81)

BEGONIACEAE (76)

Begonia L.

B. coccinea Hook. 'angel wing begonia'

Localities recorded: M(sr only)

Cultivated fibrous rooted subshrub with cane-like stems to 1.5 m, oblique glossy pale green succulent leaves to 12 cm long, dense hanging clusters of pink ♂ and ♀ flowers and fruit with three wings, pink at first then drying golden brown, shedding numerous minute brown seeds. (Other cultivated *Begonia* spp. need collecting to determine the correct names)

B. glabra Aubl. 'begonia'

Bailey

No locality recorded

Cultivated plant with ovate tooth-edged leaves and white flowers to 6 mm wide in few flowered clusters.

B. hirtella Link

Localities recorded: M(sr only)

Fibrous-rooted somewhat fleshy herb to 75 cm with asymmetrical 'pleated' leaves to 10 cm long and ♂ and ♀ greenish white to pinkish flowers in few-flowered clusters, similar to the more frequent *B. humilis* but with larger thicker more ridged leaves.

B. humilis Ait. 'begonia sauvage'

Lionnet (1)

Arc 143(K), Jef 447(S), Sch 11671,11738(K), Lam 415 (K), Fos 51968(S), Rob 2250(S)

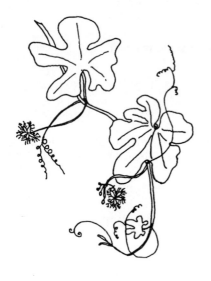

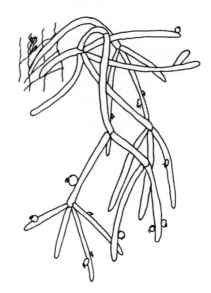

Fig. 81. Trichosanthes cucumerina   Fig. 82. Rhipsalis baccifera

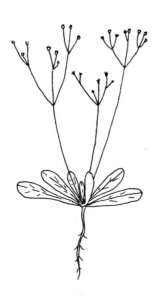

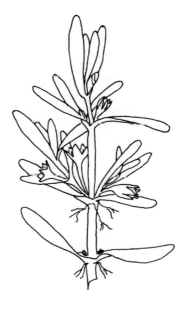

Fig. 83. Mollugo nudicaulis   Fig. 84. Sesuvium portulacastrum

Localities recorded: M

Low fleshy herb with hairy asymmetrical tooth-edged leaves to 10 cm long and axillary clusters of small white ♂ and ♀ flowers to 8 mm wide.

B. seychellensis Hemsley   'begonia sauvage, oseille marron'

Summerhayes, Bailey, Lionnet (1), Procter (1)

Jef 431,581,796(S), Gwy/Woo 1075(K), Sch 11672, 11799(K), Lam 416(K), Fos 52019(S), Ber 14664(K)

Localities recorded: M,S

Endemic fleshy herb in deep forest shade with asymmetrical leaves to 20 cm wide, deep green above and red below, with long stalked white flowers to 3 cm wide and ovoid orange fruit to 2 cm long.

B. ulmifolia Willd.

Arc 144(K), Jef 373(S), Sch 11737,11792(K), Lam 413 (K), Fos 52016(S), Rob 2423(S)

Localities recorded: M

Fleshy herb to 50 cm with bright green hairy asymmetrical leaves, longer than broad, to 15 cm long, and tightly packed clusters of many greenish pink flowers at the top of the plant.

CACTACEAE (78/1)

Hylocereus (A.Berger)Britt. & Rose

H. undatus (Haw.)Britt. & Rose   'night blooming cereus, queen of the night'

Jef sn(K)

Localities recorded: M(BG)

Cultivated epiphytic spiny climber with 3-angled stems and white flowers to 30 cm long, opening and fading in one night, with edible red fruit to 12 cm wide.

Opuntia Mill.

O. tuna (L.)Mill.   'racquette'

Bailey

No locality recorded

Shrub to 1 m with obovate joints to 12 cm or less, with pale yellow spines, flowers yellow to 5 cm wide and red edible fruit to 3 cm long. (Tuna is the mexican name for the edible fruit of flat jointed opuntias and this name is sometimes misapplied)

O. vulgaris Mill. (*O. dillenii, O. ficus-indica*)   'racquette à piquants'

Bailey, Stoddart (1)

Jef 787(S)

Localities recorded: M,Far

Erect or sprawling shrub, sometimes with a trunk, with flattened fleshy obovate spiny green joints to 30 cm long, yellow flowers to 7.5 cm wide and obovoid fleshy red edible fruit to 8 cm long.

Pereskia Mill.

P. grandifolia Haw.   'rose cactus'

Localities recorded: M(BG)(sr only)

Cultivated small tree to 5 m with spiny trunk, oblong leaves to 15 cm long, spines to 5 cm long and showy rose pink flowers to 3 cm wide and pear-shaped fruit.

Rhipsalis Gaertn.

R. baccifera (Soland. ex Mill.)Stearn (*R.cassutha*)   'mistletoe cactus, cactus'

Summerhayes, Bailey

Jef 468(S)

Localities recorded: M,S

Indigenous much branched pendant epiphytic plants, on trees or from humus in rock crevices, with slender cylindrical green stems, white flowers to 5 mm wide and pellucid white fruit to 8 mm long containing small black shiny seeds. The young plants show typical cactus bristles. (Fig. 82)

Selenicereus (A.Berger)Britt. & Rose

S. grandiflorus (L.)Britt. & Rose (*Cereus grandiflorus*)   'night flowering cactus, queen of the night'

Bailey

No locality recorded

Cultivated large creeper with stems with up to 8 rounded ribs and clusters of spines to 6 mm long, large white flowers to 22 cm long, opening at night, and fruit to 7 cm long.

AIZOACEAE (79)

Mollugo L.

M. nudicaulis Lam.

Fosberg (2)

Localities recorded: Ass

Glabrous herb with basal rosette of obovate leaves, slender erect sprays to 35 cm high of small greenish flowers and fruit a 3-valved capsule to 3 cm long containing black seeds. (Fig. 83)

M. oppositifolia L. (*Glinus oppositifolius*)

Summerhayes, Fosberg (2)

Fos 52086(S), Pro 4112,4249(K), Bru/War 106(S), Tod 13(S)

Localities recorded: D,F,A,Co,Ald

Prostrate herb with radiating stems, opposite or whorled elliptic leaves to 1.5 cm long, small green flowers, up to 4 at each node, and fruit a 3-celled capsule containing reddish brown seeds.

Sesuvium L.

S. portulacastrum (L.)L.   'pourpier'

Renvoize, Fosberg (2)

Pig sn(K)

Localities recorded: Pie,Ald,Cos,Ast

Prostrate mat-forming herb with linear or obovate fleshy leaves to 3 cm long, greenish pink solitary axillary flowers and a 3-celled capsule containing smooth black seeds. (Fig. 84)

Trianthema (Sauvag.)L.

T. portulacastrum L.

Fosberg (2)

Localities recorded: Ald

Prostrate slightly fleshy herb with unequal opposite leaves, obovate and the larger to 3 cm long, joined by the continuous petioles, solitary axillary small pinkish flowers and capsules with dehiscing lids containing small black kidney shaped seeds.

UMBELLIFERAE (80)

Apium L.

A. graveolens L.   'celery'

Bailey

Localities recorded: M(sr only), Dar(sr only)

Cultivated vegetable with thick blanched petioles and pinnately compound leaves. The petioles can be eaten raw or cooked and the leaves used as garnish.

Centella L.

C. asiatica (L.)Urb.   'bevilaqua, villaqua'

Summerhayes, Bailey, Robertson (4)

Jef 772(S), Fos 52107(S), Pro 4275,4379(K), Rob 2599(S)

Localities recorded: M,S,P,F,N,Co

Creeping herb with rounded heart-shaped leaves to 5 cm wide, shortly stalked clusters of small reddish flowers, often hidden in the leaves, and flattened small 5-ribbed fruits. It is said to have medicinal uses. (Fig. 85)

Daucus L.

D. carota L.   'carotte, carrot'

Bailey, Robertson (4)

Localities recorded: M(sr only),F

Cultivated vegetable with orange swollen edible root and feathery multifid leaves and clusters of small white flowers if not harvested.

Petroselinum Hill

P. crispum (Mill.)Nym. ex A.W.Hill   'persil, parsley'

Bailey

Localities recorded: M(sr only)

Cultivated erect glabrous short-lived perennial herb with a stout root, rosette of dark green shiny multipinnate leaves, usually very curly, with clusters of small yellowish flowers. The leaves are used as a pot herb.

Trachyspermum Link

T. copticum (L.)Link (*Carum copticum*)   'ajowan'

Bailey

No locality recorded

Annual herb to 30 cm with pinnatafid leaves and clusters of small white flowers and ovoid fruit to 1.5 mm long. The fruits yield ajowan oil which was used in the treatment of hookworm. (Bailey may have confused this with other *Carum* spp. used as herbs)

ARALIACEAE (81)

Gastonia Comm. ex Lam.

G. sechellarum (Baker)Harms (*Polyscias sechellarum*)   'bois papaye, bois banane'

Summerhayes, Bailey, Procter (1), Wilson (2)

Jef 544(S), 811(K), Pro 4095(S),4577(K), Ber 14660(K)

Localities recorded: M,S,P,F,C

Endemic thick-stemmed shrub or small tree with scaly bark, large odd-pinnate leaves to 1m long or more, the wide oblong leaflets to 30 cm long, crowded near the ends of the branches and conical tight clusters of fleshy green flowers, each to 1 cm wide.

Geopanax Hemsley

G. procumbens Hemsley

Summerhayes, Procter (1)

Ves 5538(K), Jef 792(S), Pro 4556(K)

Localities recorded: M(? no longer),S

Endemic climber with digitately compound leaves with up to 10 stalked ovate leaflets, each to 15 cm long, and spikes bearing tight clusters of fleshy greenish flowers, each to 5 mm wide.

Indokingia Hemsley

I. crassa Hemsley   'bois banane, bois de lait'

Summerhayes, Bailey, Procter (1)

Ves 5477,6138(K), Jef 464(S),486,553(K), Fos 52005, 51970(K), Ber 14630(K)

Localities recorded: M,S(? no longer),P,D,Fe

Endemic shrub or small tree with large odd-pinnate leaves with up to 9 obovate to ovate leaflets, each to 25cm long, and terminal flat-looking clusters of fleshy green flowers, each to 1.5cm wide, followed by cup-shaped fruit to 1cm long.

Polyscias J.R. & G.Forst.

P. balfouriana (Hort.Sander)L.H.Bailey (*Aralia balfouri*)   'arbuste, balfour aralia'

Bailey

No locality recorded

Cultivated shrub or small tree with speckled grey or green stems and trifoliolate leaves with heart-shaped leaflets with crenate often white margins to 10 cm wide.

P. fruticosa (L.)Harms (*Panax fruticosum*)   'arbuste'

Bailey

Localities recorded: M(sr only)

Cultivated erect shrub to 3 m with tripinnate leaves to 30 cm long, the leaflets variable and sometimes lobed with small spiny teeth, and a many-flowered terminal cluster of flowers with the ovoid fruit compressed and to 4 cm long.

P. pinnata J.R. & G.Forst.   'geranium leaf aralia'

Tod 27(S)

Localities recorded: M(sr only),F

Cultivated shrub with green and white trifoliolate or odd-pinnate leaves to 35 cm long, the leaflets circular, toothed and with white frilly margins.

Fig. 85. Centella asiatica

Fig. 86. Canthium bibracteatum

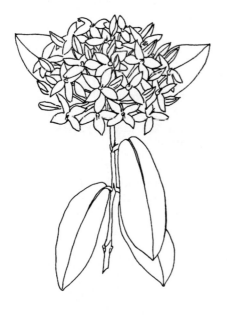

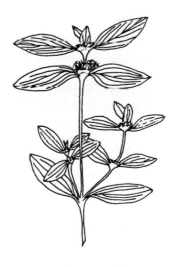

Fig. 87. Ixora chinensis

Fig. 88. Mitracarpus villosus

CAPRIFOLIACEAE (83/1)

Lonicera L.

L. japonica Thunb.   'chevre feuille, honeysuckle'

Bailey

No locality recorded

Cultivated climber with ovate leaves to 7 cm long and axillary pairs of two-lipped white or purplish fragrant flowers to 3.5cm long.

RUBIACEAE (84/1)

Amaracarpus Blume

A. pubescens Blume

Summerhayes

Jef 823(S),798(K), Pro 4371(S)

Localities recorded: M,S

Indigenous erect much branched shrub to 2 m with opposite ovate sessile leaves to 5 cm long, white flowers and pale orange or buff globose fruit to 5 mm wide.

Canthium Lam.

C. bibracteatum (Baker)Hiern   'bois dur rouge'

Summerhayes, Bailey, Piggott, Renvoize, Fosberg (2), Procter (2), Robertson (6)

Ves 5373,5537,5433(K), Arc 156(K), Jef 473,562(S), Sch 11698,11761(K), Fos 51966(S), Pro 3927,3957, 3962(S), Rob 2407,2273(S)

Localities recorded: M,S,P,C,An,Jos,Ald

Indigenous glabrous shrub, often with branches horizontal, with opposite ovate leaves to 10 cm long, with red midribs, axillary clusters of small white flowers, the petals woolly inside, and globose fruit ripening from red to black to 5 mm long. (Fig. 86)

C. carinatum Summerhayes (*C. acuminatum*)   'bois dur blanc'

Summerhayes, Bailey, Procter (1)

Pro 4070,4072,4085(S)

Localities recorded: M,S(? no longer)

Endemic erect shrub to 5 m with opposite elliptic leaves to 10 cm long, the young leaves straw coloured, few-flowered axillary clusters of small flowers on 1 cm long stalks and lobed or ridged globose fruit to 8 mm long.

C. sechellense Summerhayes

Summerhayes, Procter (1)

Arc 160(K), Pro 4078,4089,4094(S),4522(K)

Localities recorded: M,S

Endemic erect shrub to 5 m with opposite leathery ovate to rounded leaves to 15 cm long and axillary clusters of white flowers to 5 mm wide on 8 mm long stalks.

Coffea L

C. arabica L. 'cafeier, arabica coffee'

Bailey

No locality recorded

Cultivated shrub distinguished from *C. canephora* by the smaller leaves, up to 15 cm long, and by fewer fruit at each node.

C. canephora Pierre ex Froehner (*C. robusta*) 'cafeier, robusta coffee'

Bailey, Wilson (2), Robertson (4)

Fos 52130(S), Bru/War 115(S)

Localities recorded: M(sr only),P,F,A,Co

Cultivated robust glabrous shrub with erect leafy stems and horizontal fruiting stems, elliptical leaves to 20 cm long, axillary clusters of up to 6 fragrant white flowers and ovoid fruits to 1.5 cm long, ripening to red and drying black and containing 2 'beans' face to face.

C. liberica Bull ex Hiern 'cafeier, liberica coffee'

Bailey

Arc 128(K)

Localities recorded: M

Cultivated shrub similar to *C.canephora* but with larger leaves, to 30 cm long, and ovoid fruit to 3 cm long, streaked red when ripe.

Craterispermum Benth.

C. microdon Baker 'bois doux'

Summerhayes, Bailey, Procter (1), Baumer

Ves 5424,5504,5532(K), Jef 662,1191,1226(S), Fos 51994(S), 51984(K), Pro 3968,4013(S),4533(K), Ber 14626(K)

Localities recorded: M,S,P

Endemic small tree with lanceolate opposite glossy leaves to 15 cm long, a few white flowers on short flattened stems from the nodes, each flower to 1 cm long, stipules joined and rounded to 5 mm long, and red globose fruit to 5 mm long. Used medicinally in Seychelles.

Gardenia Ellis

G. augusta (L.) Merr.   'rose conde, gardenia'

Bailey

No locality recorded

Cultivated shrub to 3 m with shiny leathery opposite or whorled leaves to 7 cm long with large stipules and solitary axillary large white single or double flowers to 7 cm wide which are very fragrant. (This is often confused with *Tabernaemontana divaricata* in the Apocynaceae which is also called gardenia)

Guettarda L.

G. speciosa L.   'bois cassant bord mer'

Summerhayes, Bailey, Gwynne, Renvoize, Fosberg (1),(2), Sauer, Stoddart (1),(2),(3), Wilson (4), Robertson (4),(6)

Jef 830(S), Rob 2295,2357,2620,2851,3039,3113(S)

Localities recorded: M(sr only),S,P,F,N,C,An,L,Co,Bir,Den,Pla,Coe,Rem,Des,Dar, Jos,Poi,Mar,Alp,Ald,Cos,Ast

Indigenous pubescent shrub or tree on beach crest with horsehose-shaped leaf scars, opposite ovate or rounded leaves to 25 cm long, axillary clusters of long stalked white tubular flowers, each to 5 cm long, and globose fruit to 3 cm wide with a circular scar at the apex.

Hamelia Jacq.

H. patens Jacq.

Robertson (7)

Localities recorded: Dar

Cultivated shrub to 3 m with whorls of 3 or 4 reddish ovate leaves to 10 cm long, terminal clusters of reddish orange narrowly tubular flowers, each to 3 cm long, and ovate blackish red juicy fruits to 1 cm long.

Hedyotis L. (see also *Oldenlandia*)

H. corallicola Fosberg

Fosberg (2)

Localities recorded: Cos,Ast

Endemic prostrate mat forming herb with ovate leaves to 8 mm long with obtuse tips, small white flowers to 4 mm long, usually one per node, and capsules with black seeds.

H. lancifolia Schum. var. brevipes (Brem.)Fosberg

Fosberg (2)

Localities recorded: Ast

Slender annual herb, usually erect, with lanceolate pointed leaves to 1.5 cm long, white flowers, usually one per node and often reflexed, and capsules with brown seeds.

H. prolifera Fosberg

Fosberg (2)

Localities recorded: Ald,Ass

Endemic prostrate or ascending herb with narrow leaves to 25 cm long, small white flowers, usually 2 per node or in loose clusters, and capsule with dark brown to black seeds.

Ixora L.

I. chinensis Lam. (*I. rosea, I. stricta*)   'buisson ardent gros, ixora rose, ixora rose gros'

Bailey

No locality recorded

Cultivated shrub to 2 m with erect branches with opposite oblong leaves to 15 cm long and dense terminal clusters of red, pink, orange, yellow or white flowers, the tube to 2.5 cm long and the lobes to 7 mm long. (Fig. 87)

I. coccinea L. (*I. lutea, ?I. casei*)   'buisson ardent, ixora rouge, ixora orange, jungle flame, ?ixora grand fleur rouge'

Summerhayes, Bailey

Jef 410,589(S),1197(K), Sch 11684(K)

Localities recorded: M

Cultivated shrub to 5 m with opposite sessile leathery leaves to 9 cm and rounded clusters of red flowers to 2.5 cm wide, the tube to 3.5cm long.

I. finlaysoniana Wall. ex G.Don   'buisson'

Summerhayes, Bailey

Jef 774(S), Pro 4122(S)

Localities recorded: M

Cultivated shrub to 4 m with opposite ovate leaves to 14 cm long and terminal clusters of white flowers, each to 1 cm wide with the tube to 3 cm long.

I. hookeri (Oudem.)Bremek. (*I. odorata*)   'ixora blanc et rose'

Bailey

Dup 4/12(K), Jef 604(S)

Localities recorded: M(BG)

Cultivated shrub to 3 m with leathery opposite ovate to lanceolate leaves to 25 cm long and terminal clusters of fragrant flowers, the tube pink to 9 cm long and the petals pink outside, white inside and turning yellow with age.

I. pudica Baker   'ixora blanc'

Summerhayes, Bailey, Procter (1)

Ves 6202,6364,6445(K), Jef 516,813,1220(S), Sch 11836(K), Pro 4083,4087(S), 4538,4563(K)

Localities recorded: M,S

Endemic shrub or small tree with opposite ovate leathery leaves to 20 cm long, the young leaves purplish, terminal clusters of fragrant white flowers, with the tube to 5 mm long, the stigma exserted and the lobes reflexed, and globose fruits to 7 mm long.

Mitracarpus Zucc.

M. villosus (Sw.)DC. (*M. verticillatum*, ?*Spermacoce repens*)

Jef 359(S), Arc 141(K), Lam 405(K), Pro 3961(S), Rob 2247,2662(S), ?Sto 8093(K)

Localities recorded: M,F,?Den

Erect annual herb with ovate sessile leaves to 5 cm long and dense clusters of small sessile white flowers with finely prickly calyces in the nodes. (Fig. 88)

Morinda L.

M. citrifolia L.   'bois tortue, indian mulberry tree'

Summerhayes, Bailey, Gwynne, Renvoize, Fosberg, Stoddart (1),(2),(3), Lionnet (1), Wilson (1),(2), (4), Robertson (3),(6)

Ves 5449(K), Jef 679,810(S), Bru/War 79(S), Rob 2617,2659,2875,3098(S)

Localities recorded: M,S,P,F,N,C,A,Ma,An,L,Co,Bir,Den,Coe,Des,Dar,Poi, Mar,Alp

Indigenous shrub or small tree with glabrous opposite ovate leaves to 25 cm long, oval joined stipules, tubular white flowers to 1 cm wide on a globose spike which enlarges to form the ovoid fleshy yellow foetid fruit which tortoises love to eat.

Mussaenda L.

M. erythrophylla Schum. & Thonn.   'mussenda'

Bailey, Lionnet (2)

Localities recorded: M(BG)

Cultivated shrub with opposite petiolate ovate leaves to 14 cm long and hairy beneath, and creamy white to pink flowers in terminal clusters, each flower to 2.5 cm wide with one enlarged velvety red calyx lobe to 8 cm long.

M. frondosa L.

Jef 596(S)

Localities recorded: M(BG)

Cultivated shrub similar to *M.erythrophylla* but with the enlarged calyx lobe white and the flowers bright yellow and smaller.

Oldenlandia L. (see also *Hedyotis*)

O. corymbosa L. (*Hedyotis corymbosa*)

Summerhayes, Robertson (7)

Jef 1183(S), Fos 52139(S), Pro 4166(S), Rob 2259,2736(S)

Localities recorded: M,F,Co,Dar

Delicate annual herb (often a weed) with erect stems, opposite lanceolate leaves to 2 cm long, a few small white flowers on long stalks at each node and seed capsules to 3 mm long topped by the 4 calyx lobes.

O. goreensis (DC.) Summerhayes

Summerhayes

Pro 4527(K), Rob 3205(S)

Localities recorded: M,P

Annual prostrate or ascending herb with opposite ovate leaves to 2 cm long, terminal or apparently axillary clusters of small white or pink flowers and capsules with angular black seeds.

O. hornei Baker

Summerhayes, Procter (1)

Jef sn(K)

Localities recorded: S,Fe

Endemic herb with woody rootstock similar to *P. pentandrus* but with narrower linear leaves and fewer flowers per node, sometimes only one.

Pentas Benth.

P. bussei Krause (not *P. lanceolata*)

Jef 639(S)

Localities recorded: M(BG)

Cultivated pubescent shrub or herb with opposite ovate leaves to 10 cm long and terminal heads of bright red tubular flowers, the tube to 2 cm long.

Pentodon Hochst.

P. pentandrus (Schumach. & Thonn.)Vatke (*Oldenlandia macrophylla*)

Summerhayes

Jef 1177(S), Fea/Fra 455(S), Pro 3996,4056,4060,4114,4367(S),4459(K), Bru/War 45(S), Rob 2630(S), Tod 30(S)

Localities recorded: M,S,P,D,F,N,A,Fe,V

Annual herb with weak brittle translucent stems, opposite ovate leaves to 4 cm long, small funnel-shaped flowers to 5 mm long with white or blue 5-lobed corolla and with the 5-lobed calyx remaining on the capsule, the flowers and capsule sometimes reflexed. (Fig. 89)

Polysphaeria Hook.f.

P. multiflora Hiern   'cafe'

Renvoize, Fosberg (2)

Localities recorded: Ald,?Ass,Cos,Ast

Glabrous shrub or small tree with shreddy bark, elliptic leaves to 12 cm long, numerous small white tubular flowers in dense axillary clusters and black globose fruit to 7 mm wide.

Psychotria L.

P. dupontiae Hemsley   'bois couleuvre'

Summerhayes, Bailey, Procter (1)

Jef 415,667,723(S), Pro 3932,3976(S), Rob 2703(S)

Localities recorded: M,S

Endemic shrub to 3 m with gnarled branches, ovate sessile leaves to 9 cm long clustered at the ends of branches, terminal clusters of small almost translucent white flowers with yellow hairs in the throat and small globose ribbed fruit to 5 mm long. (Fig. 90)

P. pervillei Baker (*P. affinis, P. pallida*)   'bois couleuvre'

Summerhayes, Bailey, Procter (1), Renvoize, Fosberg (2)

Ves 5526,6137,5411,5540(K), Jef 467,797,1189(S), Pro 4011,4017,4018, 4066,4068,4093(S), Rob 2781,2809(S)

Localities recorded: M,S,P,Ald

Endemic shrub to 5 m with opposite glabrous lanceolate leaves to 15 cm long with a 1 cm long, or longer, petiole, small white flowers in long-stalked terminal clusters and globose fruit, ribbed, to 5 mm long.

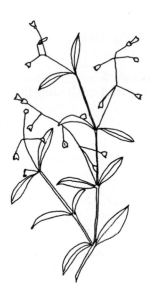

Fig. 89. Pentodon pentandrus   Fig. 90. Psychotria dupontiae

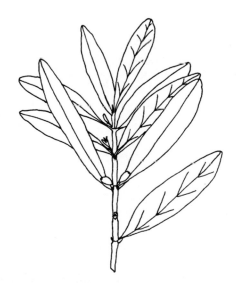

Fig. 91. Randia lancifolia   Fig. 92. Tarenna sechellensis

P. sechellarum (Baker)Summerhayes (*Psathura sechellarum*)   'bois couleuvre, bois cassant petit feuille'

Summerhayes, Bailey, Procter (1)

Ves 5527,5460,6195(K), Jef 800(S), Pro 4073,4084(S), 4553(K)

Localities recorded: M,S

Endemic shrub or small tree with leathery obovate leaves, the young leaves violet-coloured and to 20 cm long, terminal clusters of small white flowers on stout stems and globose red fruit to 8 mm long.

Randia L.

R. lancifolia (Boj. ex Baker)Hemsley   'cafe marron grand feuille'

Summerhayes, Bailey,Wilson (2)

Ves 5374(K), Arc 163(K), Jef 457,483,629,641(S), Sch 11663,11715,11752(K), Fos 51981,51982(S), Pro 3926(S), Rob 2275(S)

Localities recorded: M,S,P,Fe,C

Glabrous shrub or small tree with opposite leathery lanceolate leaves to 20 cm long, often with a reddish tinge, 1 or 2 white flowers to 1 cm long in the leaf axils and bright red fleshy oblong or ovoid fruits to 1.5 cm long. (Fig. 91)

R. sericea (Baker)Hemsley   'manglier de grand bois'

Summerhayes, Bailey, Lionnet (1), Procter (1)

Ves 5516,5529(K), Jef 768(S),761(K), Sch 11835(K), Fos 51999(S), Pro 4080(S),4515(K), Ber 14683(K)

Localities recorded: M,S

Endemic small tree with a crown of glabrous net-veined leathery ovate opposite leaves to 20 cm long with red petioles and margins, terminal clusters of tubular flowers, greyish white in bud, opening yellow and fading to red, the tube to 5 cm long, sweetly scented, and globose fruit to 6 mm long.

Rondeletia L.

R. odorata Jacq.

Dup 1/31(K)

Localities recorded: M(BG)

Cultivated shrub to 2 m with ovate dark green opposite leaves to 7 cm long and showy clusters of tubular orange red flowers with spreading rounded and slightly reflexed lobes to 1 cm wide.

Rothmannia Thunb.

R. annae (Wright)Keay (*Gardenia annae*)   'bois citron, bois calabash, wrights gardenia'

Summerhayes, Procter (1)

Pro 4101,4102,4120,4121,4135(S),4484(K), Bru/War 119(S)

Localities recorded: A,(? no longer on M,S,P)

Endemic shrub to 3 m with dark green graceful foliage, the leaves ovate to 12 cm long, large solitary axillary flowers to 4 cm long, white or white blotched with maroon, and ovoid to globose green and white mottled fruit to 5 cm long.

Tarenna Gaertn.

T. sechellensis (Baker)Summerhayes   'bois dur blanc'

Summerhayes, Procter (1)

Ves 5536,6128,6186(K), Fos 52198,52195(S), Pro 4006, 4090(S), Ber 14652(K)

Localities recorded: M,S,P,D,C,Fe

Endemic shrub to 5 m with papery ovate pointed opposite leaves to 12 cm long, terminal clusters of stalked white flowers, each to 1 cm wide and sometimes drooping, sweetly scented, and shiny globose fruit on long stalks to 7 mm wide. (Fig. 92)

T. supra-axillaris (Hemsley)Bremek.

Renvoize, Fosberg (2)

Localities recorded: Ald

Endemic shrub to 3 m with diverging branchlets, ovate opposite leaves to 10 cm long, small greenish white 5-lobed flowers in paired stalked axillary clusters and globose black fruit to 3 mm wide.

T. trichantha (Baker)Bremek.

Renvoize, Fosberg(2)

Localities recorded: Ald,Ass,Cos,Ast

Shrub or small tree with rounded ovate opposite leaves to 10 cm long, dense whitish or grey pubescent terminal clusters of white flowers and black globose fruit to 5 mm wide which can stay on the tree for several years.

T. verdcourtiana Fosberg

Renvoize, Fosberg (2)

Localities recorded: Ald,Ass

Endemic glabrous shrub with elliptic opposite leaves to 7 cm long, white 4-lobed flowers in few-flowered terminal clusters and black globose fruit to 6 mm diameter.

Timonius Rumph.

T. sechellensis Summerhayes   'bois cassant de montagne, bois cassant grand bois'

Summerhayes, Bailey, Procter (1)

Ves 5410,5535,5539,5545(K), Jef 669(S), Sch 11664, 11828, 11829(K), Fos 51972,52002(S), Pro 4540(K), Ber 14681(K), Rob 2269,2405,2808(S)

Localities recorded: M,S,P

Endemic shrub to 5 m with rhomboid opposite leaves to 15 cm long, the midrib beneath rusty, axillary few-flowered clusters of yellow flowers, the 4 corolla lobes often reflexed, and subglobose 4-angled fruits to 1 cm long, dark red when ripe. (Fig. 93)

Triainolepis Hook.f.

T. africana Hook.f. ssp. **hildebrandtii** (Vatke)Verdc.   'coeur de boeuf'

Renvoize, Fosberg (2)

Localities recorded: Ald,Ass,Cos

Much branched shrub to 3 m with elliptic leaves to 13 cm long, terminal loosely branched clusters of small fragrant white flowers, densely woolly ouside, and white to pink globose fruit to 8.5 mm wide.

Tricalysia A.Rich. ex DC.

T. sonderiana Hiern

Renvoize, Fosberg (2)

Localities recorded: Ald,Ass

Shrub or small tree to 5 m with elliptic leaves to 10 cm long, dense axillary clusters of white flowers to 1 cm long and globose black fruit to 7 mm wide.

Vangueria Juss.

V. infausta Burchell (*V. velutina*)   'vavangue'

Summerhayes, ?Bailey

No locality recorded

Similar to *V. madagascariensis* but with short stout hairs on stem and leaves.

V. madagascariensis J.F.Gmel. (*V. edulis*)   'vavangue, tamarind of the indies'

Summerhayes, Bailey

Jef 831(S), Rob 2621(S)

Localities recorded: M,S,N

Shrub to 4 m with ovate leaves to 12 cm long, dark above and pale beneath, many-flowered loose axillary clusters of greenish white flowers with reflexed corolla lobes, and globose green edible fruits to 3 cm wide with a wide circular apical scar.

COMPOSITAE (*ASTERACEAE*) (88)

Ageratum L.

A. conyzoides L.   'zerisson blanc, babouc'

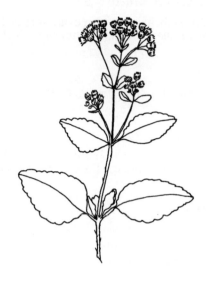

Fig. 93. Timonius sechellensis   Fig. 94. Ageratum conyzoides

Fig. 95. Bidens pilosa   Fig. 96. Coreopsis basalis

Summerhayes

Jef 357(S), Tod 53(S)

Localities recorded: M,S,F

Erect herb, a weed, with ovate serrate leaves to 9 cm long and heads of pale blue tubular florets to 5 mm wide in clusters. (Fig. 94)

Bellis L.

B. perennis L.   'marguerite, daisy'

Bailey

No locality recorded

Cultivated herb with basal rosette of obovate leaves to 5 cm long and solitary heads with white ray florets and yellow 'eye' of tubular florets to 5 cm wide.

Bidens L.

B. pilosa L.   'herbe clausette, la ville-bague, spanish needle, black jack'

Summerhayes, Bailey, Gwynne, Renvoize, Fosberg (2), Stoddart (1),(3), Wilson (4), Robertson (3)

Jef 360(S), Fra 429(S)

 Localities recorded: M,S,Rem,Des,Dar,Jos,Poi,Mar,Far,Ast

Herb to 1 m, a weed, with simple young leaves and 3–5-lobed older leaves, heads with a few white ray florets, soon falling, and yellowish brown tubular florets, the fruits linear and black to 6 mm long with up to 4 barbed bristles at the tip. (Fig. 95)

Chrysanthemum L.

C. indicum L.   'saint andre, chrysanthemum'

Bailey

Localities recorded: M(sr only)

Cultivated pubescent bedding plant to 30 cm with triangular 5-lobed and toothed leaves to 5 cm or more long, heads to 5 cm wide of yellow ray florets and tubular florets.

C. sinense Sabine   'chrysanthemum'

Bailey

No locality recorded

Cultivated herb or shrub with pubescent 5-lobed and toothed leaves to 30 cm and large colourful long stemmed flower heads, the ray florets often long and curling and white, yellow pink or mauve. Various horticultural crosses make up the garden chrysanthemum.

Cichorium L.

C. endiva L.   'chicoree, endive'

Bailey

No locality recorded

Cultivated annual salad vegetable with large tender leaves often curled at the edges and forming a 'heart'.

Coreopsis L.

C. basalis (Otto & A.Dietr.)S.F.Blake (*C.drummondii*)   'coreopsis'

Localities recorded: M(sr only)

Cultivated annual or perennial bedding plant with basal rosette of long narrowly obovate leaves or deeply divided irregularly pinnate leaves, long stalked solitary flower heads to 6 cm wide and with golden ray florets with irregularly toothed margins and brownish yellow tubular 'eye' florets. (Fig. 96)

C. tinctoria Nutt.   'coreopsis'

Bailey

No locality recorded

Cultivated annual with finely divided bipinnate leaves and erect stems bearing several flower heads to 3 cm wide with golden yellow 3-lobed ray florets with brown spots at the petal base and yellow brown tubular 'eye' florets.

Cosmos Cav.

C. bipinnatus Cav.   'cosmos, purple mexican aster'

Bailey

No locality recorded

Cultivated annual to 2 m with finely divided bipinnate leaves and flower heads to 10 cm wide with large purple, pink or white ray florets and a raised 'eye' of yellow tubular florets.

C. caudatus H.B.K.

Summerhayes

No locality recorded

Cultivated herb with bipinnatafid leaves and solitary flower heads with crimson lobed or toothed ray florets and 'eye' of yellow tubular florets, the fruit to 2 cm long with 2 long and one short barbed bristles at the tip.

C. sulphureus Cav. (*Bidens sulphurea*)

Stoddart (1)

Localities recorded: Far

Cultivated herb to 50 cm with bipinnatafid leaves and solitary flower heads with orange yellow ray florets and 'eye' of yellow tubular florets and fruit to 1.2 cm long with 2 slender yellow bristles held at right angles to the tip.

Dahlia Cav.

D. ✕ **hortensis** Guillaumin (*D. pinnata* and *D. variabilis* of gardens)

Bailey

Localities recorded: M(sr only)

Cultivated perennial bedding plant with tuberous roots, glossy bipinnate leaves to 25 cm long with ovate toothed leaflets and many types of flower heads with glossy curved petals, white, yellow, pink, red etc. (Garden dahlias are probably derived from *D. pinnata* and *D. coccinea*)

Elephantopus L.

E. mollis H.B.K. (not *E. scaber*)   'herbe la jouissance, herbe liberalis, herbe tabac'

Summerhayes, Bailey, Wilson (1), Robertson (6)

Ves 5584(K), Jef 358(S), Sch 11650(K), Fos 51963(S), Rob 2433(S)

Localities recorded: M,S,P,An

Herb to 1 m, woody at base, with a basal rosette of lanceolate, serrated edged hairy leaves and erect leafy flower spikes of heads of pale blue inconspicuous tubular florets followed by small black fruits to 4 mm long among shiny golden bracts. (Fig. 97)

Emilia Cass.

E. sonchifolia (L.)DC.

Summerhayes

Jef 497(S), Pro 3994(S), Rob 2564,2644(S)

Localities recorded: M,F,N

Erect glaucous herb to 50 cm with lobed and toothed leaves, the upper clasping the stem, and long narrow flower heads to 1.2 cm long of blue or mauve all tubular florets. (Fig. 98)

Eupatorium L.

E. ayapana Vent.   'ayapana'

Bailey

No locality recorded

Glabrous perennial herb with lanceolate 3-nerved leaves to 10 cm long with a

124

Fig. 97. Elephantopus mollis

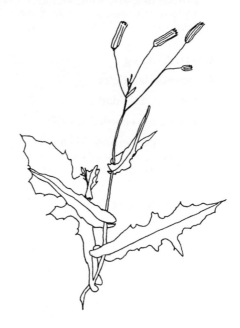

Fig. 98. Emilia sonchifolia

Fig. 99. Melanthera biflora

Fig. 100. Sigesbeckia orientalis

bronzy tint, and long stemmed clusters of many small flower heads to 7 mm wide of all tubular cream florets. Said to be used medicinally but perhaps confused with *Justicia gendarussa* in the Acanthaceae.

Gaillardia Foug.

G. lanceolata Michx.

Gwynne, Stoddart (1)

Localities recorded: Des,Far

Perennial herb to 70 cm with narrowly oblong leaves to 10 cm long, long stalked flower heads to 5 cm wide with red or yellow deeply lobed ray florets and purplish brown tubular 'eye' florets.

G. pulchella Foug. (*G.picta*)   'gaillardia'

Bailey

Pig sn(K)

Localities recorded: Pie

Cultivated bedding plant with toothed obovate or pinnately lobed leaves to 15 cm long in a basal rosette and leafy stems to 50 cm bearing solitary flower heads to 5 cm wide with ray florets with deeply incised petals, usually dark red at the base and yellow at the tip, and dark brownish red tubular 'eye' florets.

Gerbera L. ex Cass.

G. jamesonii H.Bolus ex Adlam.   'gerbera, barberton daisy'

Bailey

Localities recorded: M(sr only)

Cultivated bedding plant with a basal rosette of incised lanceolate leaves to 25 cm, woolly beneath, with solitary flower heads to 12 cm wide on leafless stalks with long narrow ray florets, sometimes in several rows, white, yellow, orange, pink or red, and yellow tubular 'eye' florets, and fruits to 6 mm long with an 8 mm long tuft.

Gynura Cass.

G. sechellensis (Baker)Hemsley (*Senecio sechellensis*)   'jacobet marron, bois chevre'

Summerhayes, Bailey, Procter (1), Baumer

Ves 6139(K), Jef 372,429(S), Sch 11673(K), Lam 414(K), Fos 51985(K), Pro 3922(S),4539(K), Ber 14638(K), Rob 2424(S)

Localities recorded: M,S

Endemic woody herb or subshrub to 1.5 m with bright green ovate succulent shiny leaves to 15 cm long and loose clusters of oblong flower heads to 1.2 cm long of orange tubular florets. Used medicinally in Seychelles.

Helianthus L.

H. annuus L.   'tournesol, sunflower'

Bailey

Localities recorded: M(sr only)

Cultivated erect hirsute annual herb to 3 m with thick angular stems, triangular to ovate leaves to 30 cm long, flower heads to 40 cm wide with a single row of yellow ray florets and many brownish tubular florets, and ovoid compressed fruit to 1.5 cm long, all white or all black or striped, containing an edible oily nut.

H. tuberosus L.   'jerusalem artichoke'

Localities recorded: M(sr only)

Cultivated perennial herb with tuberous roots, hairy stems and ovate leaves to 18 cm long, flower heads to 5 cm wide with yellow ray florets and brown tubular florets. The tuberous roots are edible and contain inulin instead of starch.

Lactuca L.

L. indica L. (*L.mauritiana*)

Summerhayes

Localities recorded: M,S

Glabrous herb to 2 m with deeply lobed sessile lanceolate leaves and narrow flower heads to 1.5 cm long with 3 rows of bracts and yellow all ray florets.

L. sativa L.   'laitue, lettuce'

Bailey

Localities recorded: M(sr only)

Cultivated annual glabrous lactiferous herb forming a dense basal rosette of crisp edible leaves and then tall branched flowering stems of flower heads to 1 cm wide of pale yellow all ray florets.

L. serricola L. (*L. scariola*)   'lastron, wild or prickly lettuce'

Bailey

No locality recorded

Annual glabrous herb with pinatafid lanceolate leaves to 18 cm long with prickly margins and flower heads to 1.5 cm wide with yellow all ray florets. (Probably an error for *Launaea intybacea*)

Launaea Cass.

L. intybacea (Jacq.)Beauverd (*Lactuca intybacea*)   'lasteron'

Renvoize, Fosberg (2), Stoddart (1)

Localities recorded: Far,Ald,Ass,Cos,Ast

Erect annual lactiferous herb with sessile lanceolate leaves to 20 cm long with toothed margins and flower heads to 1.2 cm wide of pale yellow all ray florets.

L. sarmentosa (Willd.)Alston

Renvoize, Fosberg (2)

Localities recorded: Ald,Ass,Cos

Slender prostrate perennial herb with rosettes of narrowly obovate leaves to 15 cm long and solitary flower heads to 2 cm wide of yellow all ray florets.

Melampodium L.

M. divaricatum DC.   'buton d'or'

Bailey

Dup 3/31(K), Neville 1867(K)

Localities recorded: P,D

Herb with scabrous ovate leaves and flower heads to 2.5 cm wide with broad bracts, yellow ray florets and an almost globose 'eye' of brownish yellow tubular florets.

Melanthera Rohr

M. biflora (L.)Wild

Renvoize, Fosberg (2)

Rob 3178(S), Tod 58(S)

Localities recorded: M(sr only),F,Dar,Ald,Ass

Perennial trailing cultivated and escaping herb with rough ovate toothed leaves to 7 cm long and solitary hemispherical heads to 2 cm wide of yellow ray and tubular florets. (Fig. 99)

Parthenium L.

P. hysterophorus L.   'herbe blanche'

Summerhayes, Bailey, Renvoize, Stoddart (1)

Jef 1156(S)

Localities recorded: M,S,Far

Herb with greyish white very dissected leaves and flower heads to 4 mm wide with a few white ray florets and 'eye' of yellow tubular florets.

Sigesbeckia L.

S. orientalis L.   'herbe guerit vite, herbe de flaque'

Summerhayes, Bailey, Baumer

Jef 733(S)

Localities recorded: M,S

Erect woody aromatic herb with rhomboid toothed leaves to 10 cm long and flower heads of yellow ray and tubular florets with two kinds of bracts, the outer row glandular and longer than the florets. Used medicinally in Seychelles. (Fig. 100)

Spilanthes Jacq.

S. ?mauritiana (Rich. ex Pers.)DC.

Stoddart (1)

Localities recorded: Far

Straggling hairy herb with opposite triangular ovate toothed leaves to 4 cm long and long stalked solitary axillary flower heads, cone shaped and to 7 mm long with yellow ray and tubular florets.

Synedrella Gaertn.

S. nodiflora (L.)Gaertn.

Fosberg (2), Stoddart (3)

Jef 437(S), Fos 52102(S), Pro 4152(S), Rob 2338,2579, 2672(S)

Localities recorded: M,F,N,Co,Bir,Poi,Ald

Slightly scabrid herb to 50 cm with lanceolate to ovate leaves, with irregularly toothed margins, and apparantly sessile flower heads at the nodes with few pale yellow ray and tubular florets mixed with papery bracts.

Tagetes L.

T. patula L.   'french marigold'

Stoddart (1), (2)

Localities recorded: Des,Dar

Cultivated bedding plant with pinnatafid leaves and terminal solitary flower heads to 5 cm wide with wide bright orange ray florets and 'eye' of dark brown tubular florets to 2 cm long.

Tithonia Desf. ex Juss.

T. diversifolia (Hemsley)A.Gray

Jef 750(S)

Localities recorded: M

Woody herb or subshrub to 2 m with scabrid simple or up to 5-lobed leaves to 20 cm long and long stemmed flower heads to 10 cm wide with long orange yellow ray florets and 'eye' of yellow tubular florets, the flower stem expanded below the head. (Fig. 101)

Tridax L.

T. procumbens L.   'herbe caille'

Gwynne, Renvoize, Fosberg (2), Stoddart (1),(2), Wilson (4)

Coppinger 9/1882(K), Ves 6067(K), Jef 653(S), Pro 3997(S), Fra 434(S), Rob 2320,2346,2581,3074(S)

Localities recorded: S,N,Coe,Rem,Des,Dar,Jos,Poi,Mar,Ald

Prostrate pubescent herb with some erect stems, opposite toothed ovate scabrid leaves to 6 cm long and erect long stalked solitary flower heads with pale yellow or cream ray florets which fall early and yellow tubular 'eye' florets followed by fruits to 3 mm long with a 6 mm long tuft. (Fig. 102).

Vernonia Schreb.

V. cinerea (L.)Less.   'herbe guerit vite, herbe de flaque'

Summerhayes, Bailey, Gwynne, Renvoize, Fosberg (2), Stoddart (1),(2), Robertson (6)

Jef 367(K), Pig sn,sn(K), Fos 52071(S), Pro 4451(K), Bru/War 32(S), Rob 2374,2455,2566,2676,2849,2980,3045,3092(S)

Localities recorded: M,P,F,N,A,An,L,Co,Bir,Den,Pla,Coe,Des,Dar,Jos,Poi, Alp,Pie,Far,Ald,Ass,Ast

Erect slightly pubescent annual herb, a weed, with obovate leaves with tapering bases to 7 cm long and clusters of flower heads to 5 mm wide of all tubular mauve florets and linear bracts to 4 mm long. (Fig. 103)

V. grandis (DC.)J.Humb.

Renvoize, Fosberg (2)

Localities recorded: Ald,Ass,Cos,Ast

Spreading shrub or small tree to 4 m tall with lanceolate leaves to 17 cm long and much branched clusters of flower heads to 5 mm wide of all tubular pale mauve fragrant florets.

V. sechellensis Baker   'ayapana sauvage'

Summerhayes, Bailey, Procter (1)

Localities recorded: M(?no longer)

Endemic, probably extinct, shrub to 1.5 m with tight flower heads of all tubular florets, the seed tuft pinkish and to 6 mm long.

Youngia Cass.

Y. japonica (L.)DC.

Jef 436(S), Sch 11779(K)

Localities recorded: M

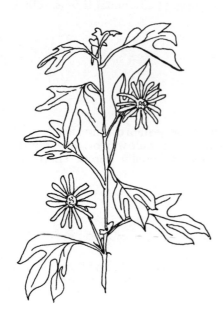

Fig. 101. Tithonia diversifolia

Fig. 102. Tridax procumbens

Fig. 103. Vernonia cinerea

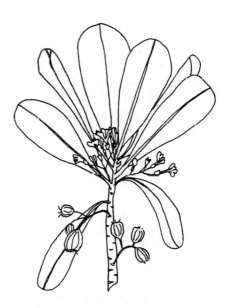

Fig. 104. Scaevola sericea

Herb to 75 cm with a basal rosette of elongated obovate toothed leaves and clusters of oblong flower heads, to 6 mm long, of all tubular florets, yellow tipped with orange.

Zinnia L.

Z. elegans Jacq.   'zinnia'

Stoddart (1)

Localities recorded: Far

Cultivated erect annual to 75 cm with opposite ovate stem clasping leaves to 10 cm long and solitary flower heads to 10 cm or more wide with bright variously coloured ray florets and hemispherical 'eye' of tubular florets. There are 'double' varieties.

Z. linearis Benth.   'zinnia'

Bailey

No locality recorded

Cultivated annual with opposite linear leaves to 6 cm long and flower heads with wide orange yellow ray florets and 'eye' of yellow tubular florets.

GOODENIACEAE (90/1)

Scaevola L.

S. sericea Vahl (*S. frutescens, S. taccada*)   'bois manioc, veloutier manioc'

Summerhayes, Sumner Gibson, Bailey, Gwynne, Renvoize, Lionnet (1), Piggott, Fosberg (2), Stoddart (1),(2), (3), Wilson (2),(3),(4), Robertson (6)

Ves 6488(K), Jef 387(S), Sch 11774(K), Fos 52047(S), Bru/War 59(S), Rob 2334,2395,2572,2653,2877,3052, 3069(S)

Localities recorded: M,P,F,N,C,A,Fe,An,L,Co,Bir,Den,Pla,Coe,Afr,Rem, Des,Dar,Jos,Poi,Mar,Def,Alp,Pro,Far,Ald,Ass,Cos,Ast

Indigenous shrub of the beach crest with glossy obovate leathery leaves to 25 cm long, white one-sided flowers to 1.5 cm long and globose to ovoid white fleshy fruit to 1.3 cm wide. (Fig. 104)

CAMPANULACEAE (91/1)

Laurentia Adans.

L. longiflora (L.)Endl. (*Isotoma longiflora, Hippobroma longiflora*)   'herbe poison, star of bethlehem'

Summerhayes, Bailey, Gwynne, Lionnet (1), Renvoize, Wilson (1), (2), Robertson (7)

Ves 6438(K), Jef 350(S), Sch 11651(K), Rob 2397, 3088(S)

Localities recorded: M,P,C,Coe,Dar

Herb to 30 cm with elongated toothed leaves, axillary white flowers with 5 lobes and a tube to 10 cm long and a fruit capsule with many small seeds. The acrid juice is poisonous. (Fig. 105)

Lobelia L.

L. chinensis Lour. (*L. radicans*)

Summerhayes

Dup 9/33(K)

Localities recorded: ?M

Low cultivated herb with slightly fleshy stems, lanceolate somewhat toothed leaves to 2 cm long and solitary axillary long-stalked pinkish purple tubular flowers to 1 cm long.

ERICACEAE (93/1)

Rhododendron L.

R. indicum (L.)Sweet   'azalea'

Bailey

No locality recorded

Cultivated shrub to 2m with lanceolate leaves to 4 cm long, hairy on both sides, and funnel-shaped rose to scarlet flowers to 5 cm wide.

PLUMBAGINACEAE (98)

Plumbago L.

P. aphylla Boj. ex Boiss.   'herbe paille en queue'

Renvoize, Fosberg (2)

Localities recorded: Ald,Ass,Cos,Ast

Almost leafless pale glaucous green woody herb or shrub to 1 m with leaves reduced to black scales and glandular clusters of white flowers, each to 15 mm long, followed by spindle shaped capsules.

P. auriculata Lam. (*P.capensis*)   'plumbago bleu'

Bailey

Jef 601(S)

Localities recorded: M

Shrub, sometimes scandent, with ribbed stems, ovate leaves to 8 cm long and clusters of pale blue flowers with the tube to 2 cm long and calyces sticky glandular towards the tip. (Fig. 106)

133

P. indica L. (*P.rosea*) 'plumbago rose'

Bailey

No locality recorded

Semi-scandent shrub with ovate leaves to 9 cm long and long spikes of rose scarlet flowers with the tube to 2.5 cm long and the calyces wholly glandular.

P. zeylanica L. 'plumbago blanc, dentelaire'

Bailey

No locality recorded

Woody creeping herb or scandent shrub with ribbed stems, ovate leaves to 13cm long and scented white flowers with the tube to 2.6cm and the calyces glandular.

MYRSINACEAE (100)

Ardisia Sw.

A. crenata Sims

Jef 395(S), Pro 3970(S), Rob 2227(S)

Localities recorded: M

Shrub to 2 m with radiating branches, ovate shiny leathery leaves to 18 cm long and with crenate edges, white flowers to 6 mm long, cone shaped in bud, and bright red globose fruit to 6 mm wide.

Rapanea Aubl.

R. seychellarum Mez

Summerhayes, Procter (1)

Pro 4079(S)

Localities recorded: M,S(? no longer)

Endemic small tree with ovate leaves to 20 cm long, paler below, and clusters of small flowers, each to 2 mm long, on the main stem or on short side shoots, and small globose fruit to 5 mm wide.

SAPOTACEAE (101/1)

Chrysophyllum L.

C. cainito L. 'star apple'

Bailey, Lionnet (2)

Jef 586(S)

Localities recorded: M(BG)

Fig. 105. Laurentia longiflora

Fig. 106. Plumbago auriculata

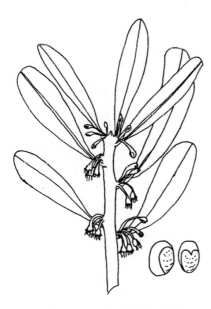

Fig. 107. Northia seychellana

Fig. 108. Maba seychellarum

Cultivated tree to 12 m with pointed ovate leaves to 15 cm long, dark green above and silky golden brown beneath, axillary clusters of small purplish white flowers, each to 4 mm wide, and green or purple globose fruit to 10 cm wide, the skin with unpleasant tasting latex but containing sweet white edible pulp and small hard brown glossy seeds.

Mimusops L.

M. coriacea (DC.)Miq. 'pomme jacko'

Bailey (not *Calophyllum sp.*)

Localities recorded: M(sr only)

Strand tree with leathery obovate leaves to 15 cm long, and long stalked globose yellow edible fruit to 4 cm wide.

M. sechellarum (Oliv.)Hemsley (*M. decipiens, M. thomassetii, Imbricaria sechellarum*) 'bois de natte'

Summerhayes, Bailey, Lionnet (1), Procter (1)

Ves 5493,6208,6374,6362,6442(K), Jef 551,568,621, 1187,1217(S), Pro 3937(S),4558(K), Ber 14690(K), Rob 2280(S)

Localities recorded: M,S,P,Fe,?Ma(Ves 5447)

Endemic tree, stunted relicts only on Mahe, with milky juice, ovate leathery leaves to 15 cm long, leaf and flower buds rusty tomentose, musky scented flowers to 1 cm wide with greenish petals from the leaf axils and globose green fruit.

Northia Hook.f.

N. seychellana Hook.f. (*N. confusa, N. hornei*) 'capucin'

Summerhayes, Bailey, Lionnet (1), Procter (1),(2)

Ves 5637,6188,6757,6324(K), Jef 1171,1230(S), Sch 11733(K), Dup 1/33,2/33,3/33,4/33(K), Pro 3978(S), 4572(K)

Localities recorded: M,S,P,C,Fe

Endemic tree to 12 m often appearing gnarled, with lanceolate leathery leaves, to 25 cm long, rusty pubescent beneath with a prominent midrib, axillary clusters of short stalked drooping greenish white musky scented flowers, each to 1.5 cm wide, and globose green fruit to 7 cm long containing one large seed, shiny brown on one side and wrinkled on the other, like a monk's face in his hood. Named after Marianne North, the 19th century painter who painted it in Seychelles and sent flowers to Kew, the first they had received. There seems to be only one species although Dupont mentions red and white varieties. (Fig. 107)

Planchonella Pierre

P. obovata (R.Br.)Pierre (*Sideroxylon ferrugineum*) 'bois mon pere, bois mozambique, bois de fer'

Summerhayes, Bailey

Ves 5249,6182,6203(K), Jef 1237(S), Pro 3950,3986, 3992,4081(S), Ber 14619(K), Rob 2542,2803(S)

Localities recorded: M,S,P

Indigenous much branched shrub or tree to 6 m with glossy ovate pale green leaves to 20 cm long, coppiced and young shoots with larger leaves, axillary clusters of small greenish white flowers, each to 5 mm wide, and pointed ovoid fleshy fruit to 1cm long.

Pouteria Aubl.

P. campechiana (H.B.K.)Baehni (*Lucuma palmeri*) 'sapote au mexique, egg fruit, yellow sapote'

Bailey

Rob 2785(S)

Localities recorded: M(BG)

Cultivated fruit tree with milky sap, long ovate wavy edged leaves to 20 cm long, many greenish yellow flowers to 8 mm long on short stalks, crowded among the terminal leaves, and egg-shaped orange yellow fruits to 7 cm long with edible mealy pulp and a large shiny brown seed.

Sideroxylon L.

S. inerme L. ssp. **cryptophlebia** (Baker)Hemsley 'bois zak, bois de natte'

Renvoize, Fosberg (2)

Localities recorded: Ald,Ass,Cos,Ast

Endemic large shrub to 5 m with obovate or rounded leaves, dense axillary clusters of very small whitish flowers, each to 3 mm long, and black globose fruit to 1.5 cm long.

EBENACEAE (102)

Diospyros L.

D. discolor Willd. 'mambolo, k k chatte, velvet apple'

Summerhayes, Bailey

Rob 2758(S), Tod 23(S)

Localities recorded: M,F

Cultivated tree to 15 m with ovate leathery leaves to 20 cm long with rusty tomentous midrib below, small silvery tomentous flowers to 1 cm wide and velvety pink fruit to 12 cm wide containing white fragrant edible flesh and large seeds.

Maba Forst. (often included in *Diospyros*)

M. seychellarum Hiern 'bois sagaye'

Summerhayes, Bailey, Procter (1)

Jef 1190(S), Sch 11719(K), Pro 3929,3949(S), 4578(K), Ber 14697(K), Rob 2277(S)

Localities recorded: M,S,P,Fe

Endemic shrub or small tree to 3 m with erect branches and glossy ovate leaves to 5 cm long in two ranks up the branches, axillary solitary white flowers to 5 mm wide ♂ and ♀ separate, and ovate orange fruit to 1 cm long, falling when ripe to leave the cup-like calyx. (Fig. 108)

STYRACACEAE (103/1)

Styrax L.

S. benzoin Dryand.   'benjoin'

Bailey

No locality recorded

Cultivated tree to 30 m with wide trunk, ovate leaves to 14 cm long, pale tomentous below, axillary sprays to 10 cm long of flowers to 1.5 cm wide, pale pubescent outside, and globose fruit to 3 cm wide. It yields gum-benzoin for medicinal uses and for incense.

OLEACEAE (104)

Jasminum L.

J. azoricum L.

Summerhayes (?may be *J. fluminense*)

Localities recorded: M

Climbing shrub with opposite trifoliolate leaves, terminal clusters of fragrant white flowers, the tube to 3 mm long and the petals to 6 mm long.

J. elegans Knobl.   'jasmin'

Renvoize, Fosberg (2)

Localities recorded: Ald

Scrambling shrub to 1m with opposite trifoliolate leaves, the longest leaflet to 5.5 cm long, clusters of fragrant white tubular flowers, the petals to 8 mm long, and globose purple black fruit to 1.2 cm wide.

J. fluminense Vell. ssp. **mauritianum** (Boj. ex DC.) Turrill (*J. mauritianum*)

Summerhayes

Localities recorded: M

Indigenous pubescent climbing shrub with opposite trifoliolate leaves, the longest leaflet to 7 cm long, with fragrant white tubular flowers in long-stalked clusters, the tube to 2.5 cm long and the petals to 1.5 cm long, and bilobed fruit to 6 mm wide.

J. multiflorum (Burm.f.)Andre   'star jasmin'

Pro 4206(S)

Localities recorded: M(sr only),D

Slender climbing pubescent shrub with opposite ovate leaves to 6 cm long with a pointed tip and dense axillary clusters of fragrant white tubular flowers, the petal lobes to 1.5 cm long.

J. sambac (L.)Ait.   'arabian jasmin'

Summerhayes

Localities recorded: M,P

Cultivated climbing pubescent shrub with opposite ovate leaves to 7 cm long and terminal clusters of fragrant white flower with the tube to 1.5 cm long and the petals to 1.5 cm long. (Probably the same as *J. muliflorum*)

Noronhia Stadm.

N. emarginata (Lam.)Poir.

Dup 217/13(K), Jef 832(S), Rob 2483(S)

Localities recorded: M,S

Coastal tree to 10 m with leathery greyish green ovate leaves to 15 cm long, indented at the tips, small fleshy yellow flowers to 8 mm wide with 4 petals and green ovate fruit to 3 cm long, said to be edible. (This is incorrectly called *Calophyllum parviflorum* in Williams, Useful and Ornamental Plants of Zanzibar and Pemba)

SALVADORACEAE (105)

Azima Lam.

A. tetracantha Lam.

Renvoize, Fosberg (2)

Jef 1162(S)

Localities recorded: M(BG),Ald.Ass.Cos,Ast

Dense shrub to 3 m with 4 spines, each to 3 cm long, at each node, ovate leathery leaves to 6 cm long, ♂ and ♀ plants with clusters of small yellowish flowers with spine tipped bracts and white berries to 8 mm wide.

Salvadora L.

S. angustifolia Turrill   'tampalocoque, tambalico'

Renvoize, Fosberg (2)

Localities recorded: Ald,Cos

Shrub or small tree to 3 m with oblanceolate strap-shaped greyish green leaves to 10 cm long, small greenish yellow flowers in pairs on short spikes and ovoid fruit to 5 mm wide.

APOCYNACEAE (106)

Allamanda L.

A. cathartica L. (incl. var. *nobilis* Moore)   'alamanda jaune, alamanda jaune grand'

Summerhayes, Bailey, Wilson (1)

Jef 591(S)

Localities recorded: M,P,L

Cultivated and escaping semi-scandent shrub with leaves in whorls of 4, ovate to 12 cm long, axillary shining large yellow bell-shaped flowers to 6 cm long (the var. to 12 cm long) and a spiny ovoid capsule containing flat brown seeds. (Fig. 109)

A. violacea Gardn. & Field   'alamanda mauve'

Bailey

Jef 634(S)

Localities recorded: M(BG)

Cultivated shrub to 2 m with arching branches, scabrid slightly hairy ovate leaves to 10 cm long, in whorls of 4, and bell-shaped dull mauve pink flowers to 6 cm wide. (Not to be confused with *Cryptostegia* in the Asclepiadaceae)

Alstonia R.Br.

A. macrophylla Wall. ex G.Don   'bois jaune'

Baumer, Wilson (1), Robertson (4),(6)

Jef 490(S), Pro 3928(S)

Localities recorded: M,P,F,An

Timber tree to 15 m with white latex, long pale green leaves to 30 cm, lax sprays of small yellowish flowers and long narrow paired pods to 40 cm long containing small flattened seeds with brown hairs at each end. Used medicinally in Seychelles.

Beaumontia Wall.

B. grandiflora (Roxb.)Wall. (*B. jerdoniana*)   'easter lily vine'

Bailey

Localities recorded: M(sr only)

Woody cultivated climber with rusty pubescent young leaves and branches, leaves to 25 cm long and axillary clusters of 5-lobed funnel-shaped white flowers to 16 cm long.

Carissa L.

C. edulis (Forssk.)Vahl (incl. var. *sechellensis*)   'bois sandal, sandal'

Summerhayes, Bailey, Procter (1), Fosberg (2)

Ves 6209(K), Fos 52199(S), Pro 4016(S),4555(K), Ber 14691(K).

Localities recorded: S(endemic var.),Ald

Small glabrous tree with opposite ovate leaves to 7 cm long, dark green above and pale beneath, axillary clusters of white fragrant tubular flowers to 1.2 cm long and globose fruit. The endemic variety is burnt on Silhouette to keep mosquitos away.

Catharanthus G.Don

C. roseus (L.)G.Don (incl. var. *alba*)   'rose amere, saponaire, pervenche, periwinkle, madagascar periwinkle'

Summerhayes, Bailey, Gwynne, Lionnet (1), Renvoize, Fosberg (2), Stoddart (1),(2),(3), Baumer, Wilson (2),(3),(4), Robertson (4),(6)

Ves 5614(K), Jef 534(S), Fos 52147(S), Fra 525(S), Bru/War 1,3(S), Rob 2315,2335,2560,2829,3010,3091(S)

Localities recorded: M,F,N,C,A,An,L,Co,Bir,Den,Pla, Coe,Rem,Des,Dar, Jos,Poi,Mar,Def,Alp,Far,Ald,Ass,Cos,Ast

Erect herb to 60 cm with obovate opposite leaves to 8 cm long, solitary or paired axillary pinkish mauve or white tubular flowers with 5 spreading free lobes to 2 cm long, and narrow fruit to 2.5 cm long containing numerous dark brown cylindrical seeds. Used medicinally in Seychelles. (Fig. 110)

Cerbera L.

C. manghas L. (*C. odollam*)   'tanghin, tanghin poison'

Summerhayes, Bailey, Wilson (2?)

Jef 1151(S), Sch 11732(K)

Localities recorded: M,S,?C

Indigenous shrub or tree to 15 m with milky latex, pointed or rounded lanceolate leaves to 20 cm long, clusters of white flowers with pink centres, the tube to 4 cm long with spreading lobes, and ovoid paired fruit to 5 cm long, often red on one side.

C. venenifera (Poir.)Steud. (*C. tanghin*)   'tanghin, ordeal plant'

Bailey

Jef 688(S), Pro 3958(S), Rob 2472(S)

Localities recorded: M

Similar to *C. manghas* with no obviously different field characters, but de L'Isle, Horne 246 and Wright 11 quoted by Summerhayes for *C. manghas* seem to belong here.

Kopsia Blume

K. arborea Blume

Pro 4284(K), Rob 2786(S)

Localities recorded: M(BG)

Cultivated spreading tree with white latex, long elliptic leaves to 20 cm, a few turning red, fragrant white flowers with the tube to 30 cm long and lobes to 3.5 cm wide, and ovoid purple fruit to 3 cm long.

K. fruticosa A.DC.   'kopsia'

Bailey

Jef 633 (S)

Localities recorded: M(BG)

Cultivated shrub with brownish black branches, pointed ovate leaves to 20 cm long and terminal clusters of pink flowers with a dark centre and white tube, lobes to 6 cm wide, the buds red.

Mascarenhasia A.DC.

M. arborescens A.DC.

Pro 4505(K)

Localities recorded: M

Small tree to 6 m with opposite dark green obovate or oblong leathery leaves, milky juice and sweet scented furry petalled pink flowers, each to 1 cm wide, in axillary clusters.

Nerium L.

N. oleander L. (*N. indicum*)   'laurier rose, oleander'

Bailey, Robertson (7)

Localities recorded: M(sr only),Dar

Cultivated shrub to 7 m with stiff lanceolate leaves to 25 cm long in whorls of 3 and pink, red or white double or single flowers to 5 cm wide, the petals asymmetrical and with a central fringe. All parts of of the plant are very poisonous. (Fig. 111)

Ochrosia Juss.

O. oppositifolia (Lam.)Schum. (*O. maculata, O. parviflora, Neiosperma oppositifolia*)   'bois chauve souris, bois jaune'

Summerhayes, Bailey, Gwynne, Renvoize, Stoddart (1),(2), Baumer, Wilson, (2),(3),(4)

Fig. 109. Allamanda cathartica

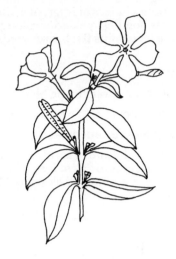

Fig. 110. Catharanthus roseus

Fig. 111. Nerium oleander

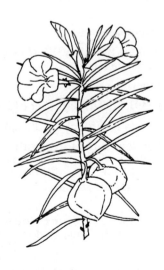

Fig. 112. Thevetia peruviana

Jef 833(S), Fos 52067(S), Rob 2840,3136,3177(S) (Kew specimens on loan so not checked)

Localities recorded: M,S,C,L,Co,Coe,Rem,Des,Dar,Mar, Def,Alp,Far

Indigenous tree with milky latex, large obovate glossy leaves to 40cm with petiole, clusters of fragrant white flowers and paired ovoid fruits to 8cm long, the yellow fleshy exterior rotting to reveal the fibrous covering to the hard nut which contains flat brown seeds. Used medicinally in Seychelles.

Plumeria L.

P. obtusa L.   'frangipane blanc'

Bailey (not *P. alba* which is a wild West Indian species), Robertson(7)

Localities recorded: M(sr only),Dar

Cultivated shrub or spreading tree to 8 m with white latex, obvate shining dark green leaves to 25 cm long, clusters of large white fragrant flowers, each to 7 cm wide, and paired elongated fruits to 20 cm long with winged seeds.

P. rubra L.   'frangipane rouge'

Bailey, Stoddart (3), Robertson (3),(4),(7)

Jef 746(S), Neville 4/1886(K)

Localities recorded: M,F,Bir,Dar,Poi

Cultivated shrub or tree to 6 m with thick cylindrical branches, white latex, elliptical pointed leaves to 25 cm long, clusters of slightly fleshy fragrant flowers to 6 cm wide, white, red, pink or yellow, with yellow centres, and long paired cylindrical fruit containing overlapping winged seeds.

Tabernaemontana L.

T. coffeoides Boj. ex A.DC. (*Conopharyngia coffeiodes, Hazunta coffeiodes*)   'bois cuiller, bois de lait'

Summerhayes, Bailey

Ves 5491,5499,6126(K), Jef 616, Pro 3987,4001(S)

Localities recorded: M,S,P,Fe,An,Cn,Ce

Indigenous shrub to 5 m with milky latex, slender branches with ovate leaves towards the tips to 10 cm long, white flowers and paired recurved pod-like fruit to 1.5 cm long with many angular whitish seeds. Wooden spoons were made from the stems.

T. divaricata (L.)R.Br. ex Roem. & Schult. (*Ervatamia coronaria, Ervatamia divaricata*) 'cafe fleur, gardenia, crape jasmin'

Bailey

Localities recorded: M(sr only)

Cultivated much branched shrub with milky sap, ovate glossy leaves to 15 cm long, waxy fragrant white flowers to 4 cm wide, often double, and paired fruit to 7 cm long with recurved beak.

T. mauritiana Poir. (*Pandaca mauritiana*)

Fosberg (2)

Localities recorded: Ald

Small tree to 5 m with milky sap, elliptic leaves to 12 cm long, at the ends of branches, clusters of creamy white flowers, the tube to 1 cm long and lobes to 2 cm wide, and paired elliptical capsules to 4.5 cm long containing many light brown ovoid and wedge-shaped seeds. (Similar to *T. coffeoides*)

Thevetia Adans.

T. peruviana (Pers.)K.Schum.   'yellow oleander, lucky nut'

Robertson (7)

Rob 3207(S)

Localities recorded: M,Dar

Cultivated shrub with milky latex, narrow leaves to 15 cm long, bright yellow or apricot flowers to 6 cm long and fleshy 4-angled fruit to 3 cm wide containing a single hard fat triangular nut. (Fig. 112)

ASCLEPIADACEAE (107)

Asclepias L.

A. curassavica L.

Summerhayes

Pro 4460(K)

Localities recorded: M,S,D

Erect herb with narrow leaves to 13 cm long, axillary clusters of red flowers to 1.2 cm wide with reflexed petals and yellow centre, and long-pointed paired fruits to 8 cm long containing flat ovate seeds each with a tuft of long white hairs. (Fig. 113)

Calotropis R.Br

C. gigantea (L.)Ait.f.

Robertson (7)

Rob 2412,2731(S)

Localities recorded: M,F,Dar

Softly woody greyish green shrub with milky latex, obovate stem clasping leaves to 15 cm long with white pubescence below, star-shaped pinkish flowers in clusters with purple centres and large inflated fruits containing seeds with a silky floss. The stem fibre can be used for fishing lines or nets. (Fig. 114)

Cryptostegia R.Br

C. madagascariensis Boj.   'rubber vine'

Rob 2732(S)

Localities recorded: F

Scandent shrub with opposite dark green glossy ovate leaves to 9 cm long, bell-shaped mauve flowers to 7 cm wide and paired woody pointed ovate fruit to 8 cm long.

Gomphocarpus R.Br.

G.fruticosus (L.)Ait.f. (*Asclepias fruticosa*)

Summerhayes

Pro 4455(K), Rob 2236(S)

Localities recorded: M,S,F

Erect woody herb or shrub with narrow leaves to 10 cm long, clusters of white and maroon flowers, and fruit inflated to an ovate ball to 7 cm long covered with soft bristles and containing seeds with tufts of hairs.

Pentopetia Decne

P. androsaemifolia Decne

Renvoize, Fosberg (2)

Localities recorded: Ass

Much branched woody shrub with opposite ovate leaves to 4 cm long, creamy white flowers to 1.2 cm wide and spindle-shaped fruit to 12 cm long containing oblong seeds with tufts of hairs.

Pleurostelma Baill.

P. cernum (Decne.)Bullock   'liane caoutchouc'

Renvoize, Fosberg (2)

Localities recorded: Ald,Ass,Cos

Low mat-forming scrambling woody herb with milky sap, oblong leaves to 3 cm long, axillary clusters of white flowers, each to 5 mm long and streaked with purple inside, and a spindle-shaped pod to 6 cm long containing ovate seeds with a tuft of hairs.

Sarcostemma R.Br.

S. viminale R.Br.   'liane sans feuilles, liane cale'

Summerhayes, Renvoize, Fosberg (2), Wilson (1)

Jef 561,618(S), Pro 3985,4003(K), Ber 14648(K), Bru/War 26(S), Rob 2742(S), Tod 9(S)

Localities recorded: M,S,P,F,C,A,Fe,Ald,Ass,Cos,Ast

Fig. 113. Asclepias curassavica   Fig. 114. Calotropis gigantea

Fig. 115. Sarcostemma viminale   Fig. 116. Heliotropium indicum

Woody, somewhat succulent climber with cylindrical green stems and leaves reduced to brownish scales, dense clusters to 3 cm wide of fragrant creamy white flowers and spindle-shaped pod (rarely found) to 12 cm long with flattened ovate seeds with apical tufts of silky hairs. (Fig. 115)

Secamone R.Br.

S. fryeri Hemsley

Renvoize, Fosberg (2)

Localities recorded: Ald,Ass,Ast

Endemic much branched climbing perennial with ovate leaves to 5 cm long, many branched clusters of white waxy flowers with lobes to 2 mm long, and paired spindle-shaped pods to 8 cm long containing ovate flattened seeds with apical tufts of silky hairs.

Toxocarpus Wight & Arn.

T. schimperianus Hemsley

Summerhayes, Procter (1)

Jef 1231(S), Pro 4219(S), Rob 2391(S), Wil/Rob 2929(S)

Localities recorded: M,C

Endemic climber with perennial rootstock and ephemeral stems, with lanceolate leaves to 15 cm long, shortly branched axillary clusters of cream or white flowers, each to 3 mm long, and paired spindle-shaped pods to 10 cm long containing ovate flattened seeds with apical tufts of silky hairs. Known recently only from Curieuse but in 1980 Wilson found many plants on Mahe in the original locality.

Tylophora R.Br.

T. indica (Burm.f.)Merr. (*T. asthmatica*) 'ipec sauvage'

Summerhayes, Bailey, Piggott, Fosberg (2)

Pro 4340,4474(K)

Localities recorded: M,Des,Mar,Ald

Trailing perennial herb with opposite ovate leaves to 6 cm long with a flexible pointed tip, axillary clusters of small green flowers on long stalks, and paired spindle-shaped pods to 9 cm long with flattened ovate seeds with apical tufts of hairs.

T. laevigata Decne.

Summerhayes

Localities recorded: S

Similar to *T. indica* but with larger leaves, different corolla-shape and clusters shortly stalked.

LOGANIACEAE (*STRYCHNACEAE*) (108)

Strychnos L.

S. spinosa Lam. (*S. annae, Brehmia spinosa*),   'calabassier du pays'

Summerhayes, Bailey, Lionnet (1), Wilson (1), Robertson (6), Baumer

Horne 330(K), Pro 4218,4280(S), Rob 2646(S)

Localities recorded: M,S,P,F,An

Indigenous shrub with paired, slightly recurved thorns, opposite obovate leaves to 6 cm long, with veins from the base, dense axillary clusters of small white flowers, and large woody globose fruit to 12 cm wide, ripening to buff and containing poisonous seeds in edible brown flesh. Used medicinally in Seychelles.

BORAGINACEAE (*CORDIACEAE, EHRETIACEAE*) (112/1)

Cordia L.

C. alliodora (Ruiz & Pav.)Oken   'laurel'

Dup 8/33(K)

Localities recorded: ?M

Tree to 20 m with slightly scabrid oblong leaves to 18cm long, clusters of many fragrant flowers, each to 1.5 cm wide and with white oblong petals.

C. myxa L.   'lacolle, glue tree'

War 191(S), Pro 4511(K), Ves 6146(K)

Localities recorded: M,A,Se

Spreading tree with drooping branches giving it a rounded look, to 6 m, with rounded leaves, slightly wavy edged to 12 cm long, terminal clusters of white flowers and globose pinkish orange fruit to 2 cm wide containing a seed embedded in glue like pulp, the fruit falling to leave a persistent chinese hat-like calyx.

C. scbestena L.   'geiger tree'

Localities recorded: M(sr only)

Cultivated tree to 10 m tall with rough textured ovate leaves to 30cm long and terminal clusters of bright orange red funnel-shaped flowers to 4cm wide with white fleshy top-shaped fruit.

C. subcordata Lam.   'porcher'

Summerhayes, Bailey, Gwynne, Sauer, Lionnet (1), Renvoize, Fosberg (2), Stoddart (1),(2),(3), Wilson (4), Robertson (3),(4),(6),(8)

Jef 533(S), Sch 11773(K), Fos 52064(S), Rob 2857,3050, 3059(S)

Localities recorded: M,P,F,N,C,A,L,Co,Bir,Den,Pla,Coe,Rem,Des,Dar,Poi, Mar,Alp,Far,Ald,Ass,Cos,Ast

Indigenous coastal tree to 10 m with broadly ovate leaves to 25 cm long and clusters of pale orange salver-shaped flowers, each to 4 cm wide, the fruit enclosed in the enlarged calyx, white at first then turning black.

Ehretia P.Br.

E. cymosa Thonn. (*E. corymbosa*)

Renvoize, Fosberg (2)

Localities recorded: Ald

Glabrous shrub to 3 m with ovate leaves to 7 cm long, flat-topped clusters of small white flowers, each to 4 mm long, and globose fruit to 6 mm wide, ripening from orange to red to bluish black.

Heliotropium L.

H. indicum L.   'herbe papillon'

Summerhayes, Bailey, Renvoize, Fosberg, Stoddart (1), Robertson (3)

Jef 502(S), Pro 4248(S), War 181(S), Rob 2735(S)

Localities recorded: M,S,F,A,Co,Poi,Far

Erect annual herb to 50 cm, a weed, with ovate sinuate edged leaves to 12 cm long and curved one-sided spikes of white or pale lavender flowers followed by the fruit, to 5 mm long, of paired beaked nutlets. (Fig. 116)

H. peruvianum L.   'heliotrope, cherry pie'

Bailey

No locality recorded

Cultivated shrub to 2 m with softly hairy branches, alternate ovate leaves and clusters of blue or purplish grey strongly scented flowers.

Tournefortia L.

T. argentea L.f.   'bois tabac, veloutier a tabac fleurs, tree heliotrope'

Summerhayes, Bailey, Gwynne, Lionnet (1), Renvoize, Fosberg (2), Stoddart (1),(2),(3), Wilson (4), Robertson(4)

Jef 1233(S), Fos 52174(S), Bru/War 85(S), Rob 2353, 2845,3032,3182(S)

Localities recorded: S,F,C,A,Co,Bir,Den,Pla,Coe,Afr, Rem,Des,Dar,Jos,Poi,Mar, Alp,Pro,Far,Ald,Ass,Cos,Ast

Indigenous shrub or small tree of beach crest with rounded crown and thick grey green furry lanceolate or obovate leaves to 30cm long, clusters of one-sided curved spikes of small white flowers, and globose fruit to 6mm wide. (Sumner Gibson gives 'veloutier' as *T. argentea* and says it grows on the coast of Mahe and on the coast and in the hills of Praslin, but this should probably be *Scaevola sericea*)

T. sarmentosa Lam.   'liane mange'

Summerhayes

Jef 571(S), Pro 4242(S)

Localities recorded: M,S

Climbing pubescent vine with harsh shiny oblong leaves to 15 cm long, sprays of paired spikes of small greenish white flowers, each to 4 mm wide, and globose fruit to 5mm wide.

Trichodesma R.Br.

T. zeylanicum (Burm.f.)R.Br.

Summerhayes

Localities recorded: M

Erect pubescent herb, a weed, with oblong leaves to 12 cm long and bristly hairy, leafy spikes of pale blue flowers, each to 1.5 cm wide and fruits surrounded by the expanded calyx.

CONVOLVULACEAE (113/1)

Argyreia Lour.

A. speciosa Sw.   'liane d'argent, elephant climber'

Bailey, Baumer

No locality recorded

Perennial woody vine with heart-shaped leaves to 30 cm long, with whitish pubescence below, flowers crowded at the ends of stalks to 30 cm long, funnel-shaped, pubescent pale lilac outside, dark purple inside, to 6 cm wide, and subglobose brown 4-seeded fruit to 1.5 cm wide. Used medicinally in Seychelles.

Cuscuta L.

C. chinensis Lam.   'cuscute, liane cuscute, liane sans feuille, dodder'

Bailey, Evans

Localities recorded: M(sr only),P

Twining leafless parasitic herb, a weed, with slender yellow stems, clusters of small whitish flowers and fruit a globose beaked capsule to 5 mm wide. (Not to be confused with *Cassytha filiformis* in the Lauraceae)

Evolvulus L.

E. alsinoides (L.)L.

Renvoize, Fosberg (2)

Localities recorded: Ald,Ass,Cos,Ast

Wiry much branched herb with ovate leaves to 8 mm long, small delicate funnel-shaped flowers to 9 mm wide, blue with a white centre and in the upper leaf axils, and globose capsules to 4 mm wide.

Ipomoea L.

I. alba L. (*Calonyction aculeatum*)   'liane manchette de la vierge, moon flower'

Summerhayes, Bailey

Localities recorded: M

Glabrous vigorous vine with heart-shaped leaves to 20 cm long, solitary white flowers to 15 cm wide, fragrant and opening at night, and ovoid pointed fruit to 3 cm long.

I. aquatica Forssk.   'bred chinois, cresson chinois, bred lamar'

Bailey

Pro 4374,4374A,4273(S), War 185(S), Rob 2604(S)

Localities recorded: M(sr only),S,N,C,A

Glabrous spreading vine on mud or in water, with dark green triangular long-stemmed leaves, long-stalked funnel-shaped flowers to 5cm wide, pale purple outside, deeper inside, and ovoid capsules to 1cm long. The leaves are eaten as spinach.

I. batatas (L.)Lam.   'patate, sweet potato'

Summerhayes, Bailey, Fosberg (2), Stoddart (3), Wilson (4), Robertson (4)

Localities recorded: M(sr only),F,Bir,Mar,Ald,Ass,Ast

Cultivated spreading prostrate vine from red, white or yellow edible tubers, with leaves variable from ovate to deeply lobed and to 15 cm long and funnel-shaped pinkish purple flowers to 5 cm wide. The young leaves are used as spinach.

I. cairica (L.)Sweet   'railway creeper'

Summerhayes

Jef 713(S), Pro 4265(S)

Localities recorded: M

Glabrous slender woody vine with 5–7-lobed palmately divided leaves and funnel shaped flowers to 6 cm long, mauve outside, deeper coloured within.

I. carnea Jacq.

Sch 11725(K), Pro 4470(K), Rob 3200(S)

Localities recorded: M

Perennial sometimes scandent shrub with triangular leaves to 10 cm long and showy clusters of pale mauve pink flowers, funnel-shaped and up to 10 cm wide. (Fig. 117)

Fig. 117. Ipomoea carnea

Fig. 118. Ipomoea pes-caprae

Fig. 119. Ipomoea quamoclit

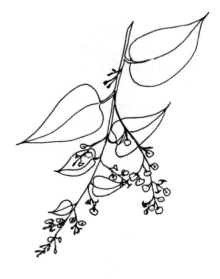

Fig. 120. Porana paniculata

I. hederifolia L. (*Quamoclit coccinea, Ipomoea coccinea*)

Summerhayes, Stoddart (3)

Jef 1211(S)

Localities recorded: M,Den

Slender annual climber with trilobed leaves and long-stalked scarlet flowers with tube to 4 cm long and spreading limb to 3 cm wide, the stamens exserted, and a round capsule to 8 cm wide containing black pubescent seeds. The name *I. coccinea* has been misapplied but that is a North American species.

I. littoralis Blume

Summerhayes

Jef 530(S), Pro 4266(S),4501(K)

Localities recorded: M,L

Indigenous trailing herb with ovate leaves to 7 cm long with deep basal lobes, and funnel shaped flowers to 5 cm long, deep purplish pink inside and paler outside. The name *I. gracilis* R.Br., which has been applied to this species in error, refers to a rare Australian endemic.

I. horsfalliae Hook.   'ipomee rouge, red clematis'

Bailey

No locality recorded

Perennial woody climber with 3–9-lobed palmately divided leaves and clusters of red bell-shaped flowers.

I. macrantha Roem. & Schultes (*I. tuba, Calonyction longiflorum, ?Calonyction tuba*)   'patatran blanc, ?convolve'

Summerhayes, Bailey, Gwynne, Renvoize, Fosberg (2), Stoddart (1),(2),(3), Wilson (2),(3),(4), Robertson (4),(6)

Jef 829(S), Fos 52049(S), Bru/War 103(S), Rob 2823, 3036,3070(S)

Localities recorded: S,F,C,A,An,Co,Bir,Den,Pla,Coe, Des,Dar,Jos,Mar, Def,Alp,Pro,Far,Ald,Ass,Cos,Ast

Indigenous vigorous glabrous vine with large heart-shaped leaves to 15 cm long, white flowers to 10 cm long and 8 cm wide, opening at night and globose capsules to 2.5 cm long containing black seeds with hairs along the edges.

I. mauritiana Jacq. (*I. digitata*)

Summerhayes

Jef 404(S), Rob 2473(S)

Localities recorded: M

Glabrous slightly fleshy vine with shallowly lobed leaves with up to 9 lobes to 18 cm long, pale beneath, funnel-shaped mauve flowers to 6 cm long in few-flowered heads and with rounded flower buds, and ovoid capsules with hairy seeds.

I. nil Roth (*I. hederacea*)

Summerhayes

Localities recorded: M

Slender twining hairy annual with shallowly lobed leaves and funnel-shaped shortly stalked flowers to 7 cm wide, deep rose pink with paler divisions and paler outside and with the calyx with long pointed lobes.

I. obscura (L.)Ker-Gawl.   'liane marron'

Summerhayes, Renvoize, Fosberg (2), Stoddart (3), Robertson (6),(7)

Jef 529(S), Pro 4247,4269(S), Rob 2437(S), Tod 10(S)

Localities recorded: M,F,An,L,Den,Dar,Ald

Slender glabrous or hairy vine with rounded leaves, heart-shaped at the base, to 7 cm long, solitary or few flowered clusters of funnel shaped creamy white or pale yellow flowers to 2.5 cm wide and with purple centres, and globose beaked capsules.

I. pes-caprae (L.)R.Br.   'patatran rouge, batata a durand, batate ronde, beach morning glory'

Summerhayes, Bailey, Gwynne, Lionnet (1), Sauer, Renvoize, Fosberg (2), Stoddart (1),(3), Wilson (2), (4), Robertson (3),(4),(6),(7)

Ves 5776(K), Jef 386(S), Fos 52161(S), Bru/War 56(S), Rob 2292, 2571,2850,3029(S)

Localities recorded: M,S,F,N,C,A,An,L,Co,Bir,Den,Pla, Rem,Des,Dar,Poi, Mar,Alp,Ald,Ass,Cos,Ast

Indigenous mat-forming creeper at beach crest with 2 lobed rounded leaves to 13 cm long, funnel-shaped mauve flowers with darker centres to 5 cm long and globose capsules to 1.8 cm wide containing blackish tomentose seeds. (Fig. 118)

I. quamoclit L.   'amourette, liane cochon, cypress vine, liane canard'

Bailey

Localities recorded: M(sr only)

Slender annual climber with deeply divided feathery green leaves and brilliant scarlet flowers with a tube to 3 cm long and acute lobes. (Fig. 119)

I. tricolor Cav. (*I. rubro-caerulea*)   'ipomee, morning glory'

Bailey

Localities recorded: M(sr only)

Cultivated glabrous annual with round to ovate leaves to 22cm long with large showy funnel-shaped sky blue flowers to 6cm long with paler tube and a 2-celled capsule with smooth black seeds.

I. venosa (Desr.)Roem. & Schultes

Summerhayes

Jef 836(S), Fos 52091,52103,52045(S), Bru/War 78(S)

Localities recorded: M,S,A,Co,Cu,Ao

Indigenous glabrous perennial with compound leaves of up to 5 narrowly ovate leaflets, clusters of large white flowers, funnel shaped and to 7.5 cm long and globose beaked capsules to 1cm long.

Merremia Dennst.

M. dissecta (Jacq.)H.G.Hall.

Summerhayes

Pro 4462(K), Ber 14636(K)

Localities recorded: M,P,D

Twiner with hairy stems, deeply palmately lobed and toothed leaves, the lobes to 9 cm long, white flowers with a red throat to 3.5 cm long and capsules enclosed in the enlarged sepals.

M. peltata (L.)Merr.   'la liane'

Summerhayes

Pro 4288(S)

Localities recorded: M,S

Indigenous vigorous vine with round peltate leaves to 30 cm long, clusters of cream or yellow funnel-shaped flowers with red spots, each to 6 cm long, and the 4 valved capsule enclosed in the calyx.

Porana Burm.f.

P. paniculata Roxb.   'liane de mai'

Dup 263(K)

Localities recorded: ?M

Strong woody pubescent climber with triangular to heart-shaped leaves to 5 cm long or more, and many-flowered white pubescent sprays of flowers each to 4 mm wide. (Fig. 120)

P. volubilis Burm.f.   'liane de mai, bridal bouquet'

Bailey

Dup 264(K)

Localities recorded: ?M

Strong woody glabrous climber with ovate leaves and large many-flowered sprays of white flowers, each to 8mm wide, the capsules enclosed by the enlarged sepals.

Stictocardia Hall.f.

S. campanulata (L.)Merr.

Summerhayes

Localities recorded: M

Indigenous woody climber with rounded or heart-shaped leaves to 10cm or more wide, rose pink flowers to 8 cm long and capsules in the expanded sepals to 4cm wide.

SOLANACEAE (114)

Atropa L.

A. belladonna L.   'belladone, deadly nightshade'

Bailey

No locality recorded

Perennial woody herb to 2 m with red sap, ovate pointed leaves to 8cm long, solitary or paired drooping dull purple flowers to 2 cm long and black glossy globose fruit to 1.2 cm wide.

Brunfelsia L.

B. australis Benth. (*B. hopeana*)   'francisea, jasmin d'afrique, yesterday today and tomorrow'

Bailey, Lionnet (2)

Jef 600(S)

Localities recorded: M

Cultivated shrub with glossy dark green oblong lanceolate leaves to 12 cm long and showy fragrant flowers to 4 cm wide, opening dark purple, and fading on successive days to pale mauve, then white.

Capsicum L.

C. annuum L. (*C. grossum*)   'piment doux, piment, tili, chillies, red or sweet pepper'

Bailey, Fosberg (2), Robertson (4)

Localities recorded; M(sr only),F,Ald

Cultivated vegetable, a herb with dark green rhomboid leaves, white flowers to 1.5 cm wide, and large green or red inflated edible fruits containing many small flat kidney-shaped white or pale yellow seeds.

C. frutescens L. (*C. minimum*)   'piment gros, piment petit, piment martin, piment arbrisseau, bird chillies, piment, tili'

Summerhayes, Bailey, Fosberg (2), Stoddart (1),(2), (3), Wilson, (1),(2),(4), Robertson (4)

Fos 52125(S), Bru/War 57(S)

Localities recorded: M(sr only),P,F,C,A,Co,Bir,Den,Rem,Dar,Jos,Mar,Ald

Cultivated herb with woody stem, sometimes zigzag, ovate leaves, small white drooping flowers and small narrow pointed fruit, hot to taste and used as a condiment or flavouring. (Fig. 121)

Cestrum L.

C. nocturnum L.   'queen of the night, night syringa, night blooming jasmine'

Bailey

No locality recorded

Cultivated shrub to 4 m with pendant branches, lanceolate or ovate leaves to 15 cm long and loose axillary clusters of narrow greenish white tubular flowers, each to 2.5 cm long, which open at night and are scented.

Datura L.

D. metel L.   'feuille de diable, fleur poison, thorn apple'

Summerhayes, Bailey, Gwynne, Lionnet (1), Fosberg (2), Stoddart (1),(3), Wilson (3),(4), Robertson (3),(5),(7)

Pig sn(K), Bru/War 5 (S), Rob 2730,3043,3142(S)

Localities recorded: F,N,A,Bir,Den,Pla,Coe,Rem,Dar,Poi,Mar,Def,Pie,Ald, Ass,Ast

Woody herb or shrub to 3 m with large triangular ovate leaves to 15 cm long with sinuate or toothed edges, large trumpet-shaped white or mauve flowers to 18 cm long, and woody reddish brown reflexed capsules covered in blunt spiny tubercles, to 4 cm long and containing whitish seeds.

D. stramonium L.   'stramonium, thorn apple'

Bailey

No locality recorded

Annual herb to 2 m with ovate coarsely toothed leaves to 18 cm long, white or violet purple 5-toothed funnel-shaped flowers to 12 cm long and erect prickly 4-valved capsules to 4.5 cm wide.

Nicotiana L.

N. tabacum L.   'tabac, tobacco'

Bailey, Fosberg (2), Stoddart (3), Wilson (3)

Bru/War 74(S)

Localities recorded: M(sr only),A,Bir,Def,Ald,Ass

Cultivated herb to 1 m with thin ovate leaves to 60 cm long, erect stems bearing many tubular pinkish flowers to 5 cm long and ovoid capsules to 15 mm long becoming 2-lobed at the top and held in the calyx with many small brown seeds. The dried and/or smoked leaves yield commercial tobacco.

Fig. 121. Capsicum frutescens

Fig. 122. Solanum anguivii

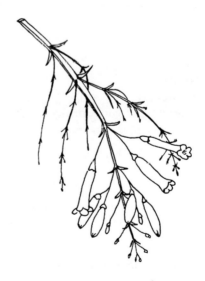

Fig. 123. Lindernia crustacea

Fig. 124. Russellia equisetiformis

Petunia Juss.

P. × hybrids Hort.   'petunia'

Bailey

No locality recorded

Cultivated glandular pubescent bedding plant with ovate leaves to 8 mm long and brightly coloured funnel-shaped velvety flowers to 7 cm long, with a deeply 5-lobed calyx, and a 2-celled capsule containing many small brown seeds.

Physalis L.

P. micrantha Link (incl. *P.angulata* L.)

Summerhayes

Pro 4176,4458(K), Rob 3145(S)

Localities recorded: S,D,F,L,Coe

Herb, a weed with triangular ovate leaves, small yellow flowers to 6 mm long and globose fruit enclosed in the papery inflated calyx which is up to 2 cm long. *P. micrantha* has yellow anthers and *P. angulata* has blue anthers.

P. peruviana L.   'groseille, pocpoc, cape gooseberry'

Summerhayes (?), Bailey, Robertson (4)

Tod (S)

Localities recorded: M(sr only),F,A

Cultivated and escaping herb with softly hairy triangular leaves to 10 cm long, yellow flowers with purple centres to 1.3 cm wide and edible globose many-seeded yellow fruit to 2 cm wide enclosed in the inflated papery calyx to 4 cm long.

Solanum L.

S. anguivii Lam. (incl. *S. indicum* var. *aldabrense* (C.H.Wright) Fosberg) (*S. indicum*) 'anguive, bringelle marron, bois zil mowa'

Bailey, Renvoize, Fosberg (2), Robertson (4)

Sch 11794(K), Pro 4473(K), Rob 2637(S)

Localities recorded: M,F,Ald

Prickly pubescent shrub to 2 m with ovate leaves to 10 cm long (less in var.) with sinuate edges, stellate hairs and prickles, white or pale mauve flowers to 1 cm wide on one-sided curved sprays, and globose red fruit to 1 cm wide. (Fig. 122)

S. lycopersicum L. (*Lycopersicon esculentum*)   'tomate, pomme d'amour, tomato'

Bailey, Fosberg (2), Robertson (4)

Localities recorded: M(sr only),F,Ald,Ass

Cultivated glandular pubescent herb with apparently pinnately compound and deeply toothed leaves, few-flowered clusters from above the leaf axils of yellow flowers to 1 cm wide and red globose edible fleshy fruit to 10 cm wide with many flattened pubescent seeds.

S. macrocarpon L.   'grosse anguive'

Bailey

Localities recorded: M(sr only)

Cultivated almost glabrous shrub with sinuate edged to lobed leaves to 18 cm long, few-flowered axillary clusters of blue flowers to 4 cm wide and whitish to pale yellow smooth globose fruit to 10 cm wide.

S. mammosum L.   'bringelle sauvage, bringelle tete, nipple fruit'

Bailey

Localities recorded: M(sr only)

Cultivated pubescent herb with deeply lobed ovate leaves to 15 cm long with spiny veins, purple flowers and large inverted pear-shaped yellow or orange fruit to 10 cm long, with nipple-shaped appendages around the base.

S. melongena L.   'bringelle, aubergine, egg plant'

Summerhayes, Bailey, Fosberg (1),(2), Stoddart (1),(3), Robertson (2),(4)

Localities recorded: M(sr only),F,Co,Bir,Coe,Far,Ald,Cos

Cultivated stellate pubescent herb to 1 m with large sinuate edged or lobed leaves to 25 cm long, purple or white flowers to 5 cm wide and large obovoid or cylindrical fruit to 20 cm or more long, picked as a vegetable when purple but ripening yellow and containing smooth whitish seeds in spongy flesh.

S. nigrum L. (incl. *S. americanum* Mill., *S. nodiflorum*)   'bred martin, black nightshade'

Summerhayes, Bailey, Gwynne, Lionnet (1), Renvoize, Fosberg (2), Stoddart (1),(2),(3), Wilson (3),(4)

Jef 522(S), Sch 11670(K), Pro 4144,4149(S), Fra 452 (S), Bru/War 37(S), Rob 2718,3015,3093(S)

Localities recorded: M,S,F,A,Bir,Den,Pla,Coe,Rem,Dar,Mar,Def,Far,Ald, Ass,Cos,Ast

Usually slender herb, a weed, with simple hairs, ovate leaves to 6 cm long, few-flowered clusters of small white flowers to 6 mm wide, nodding, and black globose fruit to 6 mm wide. The leaves can be used as spinach and medicinally. *S. americanum* is glabrous and has smaller flowers and fruit and the flower cluster is an umbel rather than a raceme.

S. seaforthianum Andr.   'potato creeper'

Bailey

Localities recorded: M(sr only)

Ornamental perennial climber with dark green deeply pinnatafid leaves to 18 cm long, clusters of lilac blue flowers, each to 4 cm wide with protruding yellow stamens, and red globose fruit.

S. tuberosum L.   'pomme de terre, potato'

Bailey

Localities recorded: M(sr only)

Cultivated herb with deeply pinnately lobed leaves and white, red or purple flowers with small inedible yellowish globose fruit to 2 cm wide, grown for its edible starchy tubers.

SCROPHULARIACEAE (115)

Angelonia Benth.
A. biflora Benth.

Summerhayes

Localities recorded: M

Ornamental herb to 1 m with glandular lanceolate toothed leaves to 10 cm long and dark blue flowers in the leaf axils, opening up the spike, each to 1.5 cm long. There are pink and white varieties.

Antirrhinum L.
A. majus L.   'gueule de loup, snap dragon'

Bailey

Localities recorded: M(sr only)

Ornamental herb with lanceolate leaves to 7 cm, spikes of large colourful tubular two-lipped flowers, each to 5 cm long, and conical capsules with the persistent style and many small brown seeds.

Bacopa Aubl.
B. monnieri (L.)Pennell

Renvoize, Fosberg (2)

Localities recorded: Ald

Small aquatic or terrestrial herb with sessile obovate leaves to 1 cm long, solitary axillary bell shaped white or pale pink or blue flowers to 8 mm long and ovoid capsules to 8 mm long.

Bryodes Benth.
B. micrantha Benth.

Fosberg (2)

Localities recorded: Ald

Inconspicuous moss-like herb with sessile linear leaves to 3.5mm long, tiny white or pale mauve flowers and a globose capsule to 1.2 mm wide.

Lindernia All.

L. crustacea (L.)F.Muell.

Summerhayes

Jef 752(S)

Localities recorded: M

Low creeping herb with triangular ovate toothed opposite leaves to 8 mm long, blue mauve flowers to 6 mm long on stalks to 1cm long and capsules with many small seeds. (Fig. 123)

Russellia Jacq.

R. equisetiformis Cham. & Schlect.   'goutte de sang, coral plant'

Summerhayes, Bailey, Robertson (7)

Localities recorded: M,Dar

Ornamental herb or subshrub with many arching green ridged stems, often rooting at the tips, lanceolate leaves to 1.5 cm long and narrow bell-shaped red flowers to 2.5 cm long on slender stalks. (Fig. 124)

Scoparia L.

S. dulcis L.

Summerhayes

Arc 130(K), Jef 710(S), Sch 11778(K), Pro 4130,4262 (S), Rob 2708,2712(S)

Localities recorded: M,S,P,F

Much branched small woody herb, a weed, with opposite ovate toothed leaves to 3.5 cm long, white flowers to 5 mm wide on stalks to 1cm long, and globose capsules to 3 mm wide.

Striga Lour.

S. asiatica (L.)Kuntze   'herb de feu, herbe rouge, herbe de riz'

Summerhayes, Bailey, Gwynne, Lionnet (1), Renvoize, Fosberg (2), Stoddart (1),(2),(3), Baumer, Wilson (2), Robertson (6)

Jef 389(S), Sch 11697(K), Pro 4163(S),4262(K), Rob 2348,2438,2733, 2859,2972,3031,3062(S)

Localities recorded: M,S,P,D,F,C,An,Bir,Den,Pla,Coe,Des,Dar,Jos,Poi,Alp, Pro,Far,Ald,Cos

Scabrid parasitic (on Gramineae) herb, slender and erect to 40 cm, with linear leaves to 3 cm long, scarlet or yellow irregular flowers to 1 cm long and ovoid capsules to 6 mm long enclosed in the persistent calyx. Used medicinally in Seychelles. (Fig. 125)

S. hermonthica Benth.

Bailey

No locality recorded

Parasitic scabrid herb similar to *S. asiatica* but larger and with mauve flowers.

Torenia L.

T. fournieri Linden. ex Fourn.   'pennsee malgache'

Bailey

No locality recorded

Ornamental bushy herb to 30 cm with square stems, opposite ovate tooth-edged leaves to 5 cm long and attractive funnel-shaped bluish purple flowers with a yellow spot in the throat, to 3 cm long.

GESNERIACEAE (119)

Episcia Mart.

E. cupreata (Hook.)Hanst.   'flame violet'

Localities recorded: M(sr only)

Ornamental pot plant with variegated hairy silvery and green ovate leaves to 12 cm long and scarlet tubular flowers with spreading petals to 2 cm wide, the flower stalks elongating to bear new plantlets.

E. reptans Mart.

Localities recorded: M(sr only)

Ornamental pot plant similar to *E. cupreata* but with darker more heavily embossed leaves and darker red flowers.

Saintpaulia H.Wendl.

S. spp.   'african violet'

Localities recorded: M(sr only)

Ornamental pot plants with leaves mostly basal, ovate to rounded, pubescent and slightly fleshy, with variously coloured in shades of white, pink or blue, single or double flowers to 3 cm wide, and rounded or conical green capsules.

BIGNONIACEAE (120/1)

Colea Boj. ex Meissn.

C. seychellarum Seem. (*C. pedunculata*)   'bilimbi marron'

Fig. 125. Striga asiatica

Fig. 126. Saritaea magnifica

Fig. 127. Tabebuia pallida

Fig. 128. Barleria prionitis

Summerhayes, Bailey, Procter (1), ?Evans

Jef 466(S),820(K), Sch 11659(K), Fos 52018(S),51962(K), Pro 3955, 4516,4562(K), Ber 14667(K), Rob 2638, 2768(S)

Localities recorded: M,S,?P

Endemic tree to 6 m with odd pinnate leaves to 40 cm long with ovate leaflets to 12 cm long, few-flowered clusters of tubular yellow flowers to 4 cm long and with maroon markings which arise from the main trunk and branches, and cylindrical fruit to 10 cm long.

Crescentia L.

C. cujete L.   'calabassier'

Bailey, Lionnet (2)

Localities recorded: M(BG)

Cultivated small tree to 8 m with spreading branches bearing tufts of oblanceolate leaves to 20 cm long, yellowish tubular flowers with mauve markings and globose fruit to 30 cm wide which has hard skin when ripe and can be used as a utensil or for musical instruments. Bailey refers to another unnamed species also as 'calabassier' and this may be *C. alata* H.B.K. which has trifoliolate leaves and cream and red flowers but there is no record.

Jacaranda Juss.

J. mimosifolia D.Don

Bailey (not *J. micrantha*)

Localities recorded: M(sr only)

Ornamental tree to 20 m with feathery bipinnate leaves with many pointed ovate leaflets each to 1 cm long, sprays of bluish mauve tubular flowers, each to 5 cm long, and round flat fruit to 6 cm wide and woody.

Millingtonia L.f.

M. hortensis L.f.   'indian cork tree'

Localities recorded: M(sr only)

Ornamental tree to 30 m with corky bark, opposite bi- or tripinnate leaves to 50 cm long, showy white long-tubed fragrant flowers to 7 cm long and a narrow two-valved capsule to 33 cm long.

Saritaea Dugand.

S. magnifica (Sprague ex van Steenis)Dugand.

Jef 595(S)

Localities recorded: M(BG)

Ornamental climber with shiny leathery opposite leaves, each with two obovate leaflets to 10 cm long and sometimes a tendril, showy funnel-shaped purplish pink flowers, to 6 cm wide, the tube white with purple markings, and a linear capsule to 30 cm long. (Fig. 126)

Spathodea P. Beauv.

S. campanulata P. Beauv. (*S. nilotica*)

Bailey, Lionnet (1), Wilson (1)

Localities recorded: M(sr only),P

Cultivated and escaping tree to 20 m with pale grey bark, opposite odd-pinnate leaves to 50 cm long, showy orange red cup-shaped flowers edged with yellow to 10 cm long, the buds tomentous and full of water which can be squirted out, and hard blackish brown capsules to 25 cm long containing winged seeds.

Tabebuia Gomez ex DC.

T. pallida (Lindl.)Miers (*T. heterophylla, T. leucoxylon*) 'calice du pape, tecoma, white cedar'

Sumner Gibson, Bailey, Lionnet (1), Procter (2), Fosberg (2), Stoddart (1),(3), Wilson (2), Robertson (4),(6),(7)

Jef 442(S), Rob 3153(S)

Localities recorded: M,S,P,F,C,An,Bir,Den,Coe,Dar,Far,Ast

Cultivated and escaping tree to 20 m with 3–5-foliolate leaves with obovate leaflets to 10 cm long, solitary or few-flowered clusters of pale pinkish mauve tubular flowers with spreading lobes to 7 cm long and narrow cylindrical capsules to 18 cm long containing flat ovate seeds with a wing at each end. (Fig. 127)

T. rosea (Bertol.)DC. (*T. pentaphylla*) 'calice du pape, tecoma rose, pink poui'

Bailey, Lionnet (2)

Localities recorded: M(BG)

Cultivated tree similar to *T. pallida* but with the leaflets elliptic and pointed, several flowers per cluster, pink with yellow throat, and longer pods, to 30 cm long.

Tecoma Juss.

T. stans (L.)H.B.K. 'tecoma a fleurs jaunes'

Bailey, Robertson (5)

Jef 592(S)

Localities recorded: M(BG),N

Ornamental shrub to 4 m with opposite odd-pinnate leaves with elliptic serrated edged leaflets to 10 cm long, terminal sprays of yellow trumpet-shaped flowers, each to 6 cm long, and narrow capsules to 25 cm long containing many flat brown winged seeds.

ACANTHACEAE (122)

Asystasia Blume

A. gangetica (L.)T.Anders. (*A. bojeri, A. multifora*)   'herbe mange tout, mange tout, coromandel grass'

Summerhayes, Bailey, Gwynne, Renvoize, Fosberg (2), Stoddart (1),(2),(3), Wilson (2),(4), Robertson (6)

Dup 7/33(K), Ves 5162,6439(K), Jef 355(S), Pig sn,sn(K), Sch 11743(K), Fos 52041,52153,52166,51991(S), Rob 2257,2577,2681,2968,3155(S),2756(K)

Localities recorded: M,P,F,N,C,A,An,Co,Cu,Bir,Den,Coe,Dar,Mar,Pro,Pie, Far,Ald,Cos

Softly hairy herb, rooting at the nodes with opposite ovate pointed leaves to 10 cm long, slender spikes of tubular two-lipped white flowers to 1 cm long with green marks in the throat, opening one or two at a time, and compressed club-shaped pointed capsules to 2 cm long containing flat brown seeds.

Barleria L.

B. cristata L.

Summerhayes

Localities recorded: M

Cultivated hairy shrub to 2 m with opposite ovate leaves to 12 cm long, axillary sessile clusters of purplish or white tubular flowers each to 7 cm long with prickly sepals and club shaped capsules. Used as a hedge plant.

B. decaisniana Nees

Renvoize, Fosberg (2)

Localities recorded: Ast,(not on Ald)

Shrub to 60 cm with opposite ovate leaves to 10 cm long, one-sided clusters of blue trumpet-shaped flowers, to 3.5 cm long and oblong capsules to 1.5 cm long, blackish.

B. prionitis L.

Summerhayes

Jef 771(S), Pro 4472(K)

Localities recorded: M,P

Shrub to 2 m with opposite ovate leaves to 10 cm long, slender grey axillary spines to 2 cm long, yellow 5-lobed tubular flowers to 4 cm long and capsule beaked and club-shaped to 1.5 cm long. (Fig. 128)

Crossandra Salisb.

C. infundibuliformis (L.)Nees (*C. undulaefolia*)   'crossandra'

Bailey

Jef 635(S)

Localities recorded: M

Ornamental woody herb with pointed ovate dark green leaves to 12 cm long, erect spikes of salmon pink flowers, usually several open at once, each to 4 cm wide, and oblong capsules to 1 cm long with seeds with fringed scales.

Graptophyllum Nees

G. pictum (L.)Griff.

Summerhayes

Localities recorded: M

Ornamental shrub to 4 m with opposite ovate leaves to 15 cm long, variously coloured, green with white spots, reddish brown or green with a yellow centre and red leaf stalk, with short erect spikes of dark purple two-lipped flowers to 4 cm long, a few opening at a time.

Hemigraphis Nees

H. alternata (Burm.f.)T.Anders. (*H. colorata*)   'red ivy'

Rob 3208(S)

Localities recorded: M

Ornamental perennial herb, rooting at the nodes, with dark purple ovate sinuate edged leaves to 6 cm long with a silvery grey sheen above, and short spikes of tubular white flowers to 1.5 cm long, a few opening at a time with overlapping bracts between them.

Hypoestes R.Br.

H. aldabrensis Baker

Renvoize, Fosberg (2)

Localities recorded: Ald,Ass,Cos,Ast

Endemic perennial herb to 20 cm tall with ovate opposite leaves to 3 cm long, blue or lilac flowers in leafy spikes with papery bracts, and club-shaped capsules to 8 mm long.

Justicia L.

J. betonica L.

Rob 3180(S)

Localities recorded: Dar

Introduced weed, a woody herb to 1.5 m with opposite narrowly elliptic leaves, terminal spikes of white two-lipped flowers to 1.5 cm long with mauve marks in the throat, held in leafy bracts with marked veination, and club-shaped pale brown capsules to 1.6 cm long.

J. brasiliana Roth (*Beloperone amherstiae*)

Jef 603(S)

Localities recorded: M(BG)

Ornamental sub-shrub to 1m with arching branches, opposite narrow ovate leaves to 6 cm long, clusters of pink flowers each to 4cm long at the nodes and club-shaped dark brown capsules to 1.5 cm long.

J. gardineri Turrill

Summerhayes, Procter (1)

Localities recorded: S

Endemic (?), woody herb with opposite ovate leaves to 15 cm long, petiole to 4 cm long, short axillary spikes of (?) white flowers, each to 4 mm long and club-shaped shiny capsules to 8 mm long.

J. gendarussa Burm.f.   'ayapana sauvage, natchouli, lapsouli'

Summerhayes, Bailey, Wilson (1)

Jef 773(S), Sch 11708(K), Fos 52088(S),52026(K), Pro 3925(S), Rob 2270(S)

Localities recorded: M,P,Co

Indigenous shrubby herb to 2 m with purple stems and narrowly elliptical leaves to 15 cm long with purple midrib and often purple tinged blade, and short terminal spike of tubular two-lipped white flowers with purple marks, each to 1.5 cm long. (Fig. 129)

J. procumbens L.

Fosberg (2)

Localities recorded: Ald,Ass

Trailing annual herb with opposite oblong leaves to 2.5 cm long, short terminal spikes of flowers and sharp bracts and ovoid capsules to 3 mm long.

Pseuderanthemum Radlk.

P. carruthersii (Seem.)Guill. (*P. atropurpureum, Eranthemum tricolor*)   'tricolore'

Summerhayes, Bailey

Rob 2706(K)

Localities recorded: M

Ornamental shrub with variegated olive green and yellow, or reddish purple and grey, leaves with sinuate margins to 10 cm long, or ovate golden yellow leaves to 18 cm long, short terminal spikes of white flowers to 1.5 cm long with mauve marks in the throat and club-shaped capsules to 1.5 cm long.

P. malabaricum (C.B.Cl.)Gamble

Summerhayes

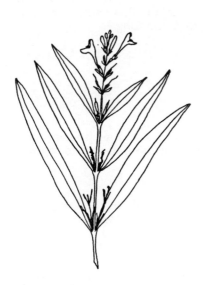

Fig. 129. Justicia gendarussa

Fig. 130. Thunbergia grandiflora

Fig. 131. Clerodendrum thomsonae

Fig. 132. Duranta erecta

Localities recorded: M

Ornamental shrub with opposite ovate leaves to 15 cm long, long slender terminal spikes of tubular light mauve flowers, the tube narrow to 2 cm long and the regularly 5-lobed mouth to 1.5 cm wide, the flowers in spaced out clusters, and pointed club-shaped capsules to 2.5 cm long.

Ruellia L.

R. monanthos (Nees)Boj.

Fosberg (2)

Localities recorded: Ald,Ass

Small perennial herb with broad ovate leaves to 1.5 cm long, solitary white and mauve trumpet-shaped flowers to 1.5 cm long and club-shaped capsules to 1 cm long.

Strobilanthes Blume

S.sp.

Rob 2757(S)

Localities recorded: M

Ornamental shrub with opposite ovate leaves to 15 cm with well marked veins, stout leafy spikes of trumpet-shaped pale mauve flowers with yellow marks inside the lower lip, each to 2 cm long and narrow stiff capsules, purple when ripe, to 4 cm long.

Thunbergia Retz.

T. alata Boj. ex Hook.   'black eyed susan'

Summerhayes, Bailey

Localities recorded: M

Perennial twining herb with triangular leaves to 7 cm long with a winged petiole and tubular flowers with spreading lobes to 4 cm wide, usually orange with black inside the throat.

T. erecta (Benth.)T. Anders.   'thunbergia'

Summerhayes, Bailey, Robertson (7)

Localities recorded: M,S,Dar

Ornamental shrub to 2 m with brittle square twigs, opposite ovate leaves to 6 cm long and large solitary stalked trumpet-shaped flowers with yellow throat and spreading dark or pale blue lobes to 6 cm wide, the white paired bracts falling as the flower opens. There is also a white flowered variety.

T. fragrans Roxb.

Jef 480(S)

Localities recorded: M

Trailing perennial herb with triangular leaves with two short basal lobes to 7 cm long and white tubular flowers with spreading lobes to 6 cm wide.

T. grandiflora Roxb.   'blue trumpet vine'

Jef 368(S)

Localities recorded: M

Vigorous ornamental climber with ovate to triangular somewhat toothed leaves to 15 cm long and 3–5-nerved from the base, slightly scabrid or hairy, and large trumpet shaped sky blue or white flowers with the spreading lobes to 8 cm wide. (Fig. 130) (*T. laurifolia* Lindl. has glabrous leaves with only 3 veins)

VERBENACEAE (125/1)

Clerodendrum L.

C. ?calamitosum L.

Lam 405(K)

Localities recorded: M

Shrub with opposite irregularly toothed leaves, terminal spikes of tubular white flowers, the tube to 2 cm long and the spreading lobes to 1.5 cm wide, with the papery calyx expanded to hold the globose black fruit which are 6 mm wide.

C. fragrans (Vent.)Willd.   'modestie'

Summerhayes, Bailey, Lionnet (1)

Sch 11727(K), Rob 2436(S)

Localities recorded: M

Erect pubescent subshrub with large triangular somewhat toothed leaves to 25 cm long and dense heads of double pale pink or white flowers with dark pinkish red sepals, each to 3 cm wide.

C. glabrum E.H.Meyer

Renvoize, Fosberg (2)

Localities recorded: Ald,Ass,Cos,Ast

Endemic shrub or small tree to 5 m with ovate leaves to 9 cm long, terminal clusters of white salver-shaped flowers each to 9 mm long, and globose orange fruit to 6 mm wide.

C. serratum Spreng.

Summerhayes

Pro sn(S)

Localities recorded: M

Woody herb or subshrub to 2 m with ovate serrated edged leaves to 20 cm long, stout terminal spikes of blue flowers each to 1.2 cm long with long stamens and 2-, 3- or 4-lobed black fruit to 8 mm wide.

C. speciosissimum van Gecrt ex C.F.A. Morren (*C. fallax*) 'modestie rouge, glorybower'

Summerhayes, Bailey, Lionnet (1),Baumer, Wilson (1)

Jef 433,582(S), Sch 11658(K), Pro 3924(S), Rob 2422(S)

Localities recorded: M,P

Erect or spreading softly pubescent shrub to 4 m with heart-shaped leaves to 25 cm long, terminal clusters of scarlet flowers with red sepals and stems, each to 1.5 cm wide and 4-lobed fruit, green at first, then black, held in the red sepals. Used medicinally in Seychelles.

C. thomsonae Balf.f. 'bleeding heart vine, coeur de jesus'

Summerhayes, Bailey (not his *C.ugandense*)

Localities recorded: M

Ornamental climber with ovate dark veined leaves to 12 cm long, and sprays of flowers with white sepals and blood-red petals with exserted stamens each to 3 cm long. (Fig. 131)

Congea Roxb.

C. griffithiana Munir (*C. tomentosa*) 'congea'

Bailey, Lionnet (2), Fosberg (2)

Jef 747(S), Sch 11815(K)

Localities recorded: M(BG),Ald

Ornamental pubescent climber with ovate leaves to 12 cm long and loose sprays of clusters of small hairy flowers, each cluster backed by 3 or 4 velvety pale mauve bracts, each bract to 3 cm long.

Duranta L.

D. erecta L. (*D. repens*)

Summerhayes, Robertson (7)

Localities recorded: M,Dar

Ornamental shrub with square slender drooping branches with opposite ovate serrated edged leaves to 7 cm long, sprays of white or mauve 5-lobed tubular flowers, each to 1.5 cm wide fragrant at night, and globose shiny orange fruit to 7 mm wide. (Fig. 132)

Gmelina L.

G. phillipensis Cham. (*G. hystrix*) 'hedgehog'

Arc 153 (K), Jef sn(K)

Localities recorded: M

Ornamental spiny climbing shrub with elliptic leaves to 8 cm long and large yellow bell-shaped 2-lipped flowers to 8 cm long with purplish bracts to 5 cm long.

Karomia Dop.

K. speciosa (Hutch. & Corb.) R. Fernandes (*Holmskioldia speciosa*)

Jef 602(S)

Localities recorded: M

Ornamental shrub with opposite triangular toothed leaves to 5 cm long and clusters of deep blue flowers in spreading funnel-shaped persistent pink calyces.

Lantana L.

L. camara L.   'lantana, vieille fille, lerbe soulie'

Summerhayes, Sumner Gibson, Bailey, Lionnet (1), Renvoize, Fosberg (2), Wilson (1), Robertson (6)

Jef 441(S), Rob 2429,3124(S)

Localities recorded: M,S,P,An,L,Coe,Ald

Spreading shrub, usually with square prickly stems, opposite ovate tooth edged leaves to 7 cm long, rough on both surfaces, flat topped heads of small tubular flowers with spreading lobes, each to 9 mm wide, variously coloured, orange, pink or white with the throat yellow,and deep purple globose fruit to 5 mm wide. (Fig. 133)

Lippia L.

L. nodiflora (L.) Rich. (*Phyla nodiflora*)   'gazon verveine, verveine, lippia'

Summerhayes, Bailey, Gwynne, Renvoize, Stoddart (1),(2),(3), Baumer, Wilson (4)

Ves 5657(K), Jef 658(S), Pig sn(K), Arc 163(K), Pro 4009,4042(S),4295(K), Fra 440(S), Rob 2291,2607,2870, 2951,3001,2378,3075(S), Tod 47(S)

Localities recorded: M,S,P,F,N,Bir,Den,Pla,Coe,Rem,Des,Dar,Poi,Mar,Alp, Pie,Far

Indigenous creeping mat-forming herb with obovate serrated edged opposite leaves to 4 cm long and small cylindrical heads, to 1.5 cm long, of small mauve and white flowers, one ring opening at a time. Used medicinally in Seychelles. (Fig. 134)

Petrea L.

P. volubilis L.   'petrea'

Bailey

Localities recorded: M(sr only)

Ornamental climbing shrub with opposite brittle rough ovate leaves to 15 cm long and sprays of flowers with pale purple persistent calyces with 5 long lobes, each to 2 cm long, and corolla of dark purple falling early.

Premna L.

P. obtusifolia R.Br. (*P. corymbosa*) 'bois sureau, bois sirop'

Summerhayes, Bailey, Renvoize, Fosberg (2), Wilson (2), Robertson (6),(7)

Ves 5445,5491A(K), Arc 148(K), Jef 374(S), Fos 52194 (S), Pro 3988(S), Bru/War 76(S), War 184(S), Rob 2474,2655(S), Rob/Wil 2776,2783(S), Rob/Gre 2790(S)

Localities recorded: M,S,P,F,C,A,An,Cn,Dar,Ald,Ass,Cos,Ast

Indigenous shrub or small tree with opposite ovate leaves to 15 cm long and petioles to 5 cm long, untidy heads of small greenish flowers, each to 5 mm wide, and globose black fruits to 5 mm wide.

Stachytarpheta Vahl

S. jamaicensis (L.)Vahl 'epi-bleu queue de rat, verveine bleu, jamaican verveine'

Summerhayes (his *S.indica* in part, specimens quoted), Bailey, Gwynne, Renvoize, Sauer, Fosberg (2), Stoddart (1),(2),(3), Lionnet (1), Wilson (1),(3),(4), Robertson (6)

Thomasset 39, Gardiner 5684A, Fryer n/n, Gardiner 50, Horne 552 — all at Kew, Ves 5943(K), Jef 500(S), Fos 52074(S), Fra 448(S), Rob 2286,2314,2369,2561,2687,2856,2955,3049,3081(S)

Localities recorded: M,P,F,N,C,An,L,Co,Bir,Den,Pla,Coe,Rem,Des,Dar,Jos, Poi,Mar,Def,Alp,Pie,Far,Ald, Ass,Ast

Woody herb with low spreading branches, opposite ovate leaves with broad blunt teeth and rounded tip to 10 cm long, often pale green, and long spikes to 5 mm thick with pale blue flowers, a few opening at a time.

S. mutabilis (Jacq.)Vahl 'epi-bleu rose, changeable verveine'

Summerhayes, Bailey, Lionnet (1)

Jef 780(S), Fra 421(S), Pro 4445(K), Rob 2430(S)

Localities recorded: M

Erect woody herb to 1.5 m with opposite ovate serrated edged pubescent leaves to 12 cm long and long spikes to 8 mm wide of brick red or pink flowers, a few opening at a time.

S. urticifolia Sims (*S. indica*) 'epi-bleu, nettle leaved verveine'

Summerhayes (his *S. indica* in part, specimens quoted), Bailey, Lionnet (1), Sauer, Fosberg (2), Wilson (1), Robertson (6)

Thomasset 40, Gardiner 142,5684B, Horne 590, Kirk n/n — all at Kew, Arc

Fig. 133. Lantana camara

Fig. 134. Lippia nodiflora

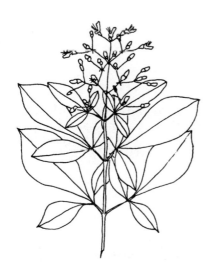

Fig. 135. Vitex trifolia

Fig. 136. Avicennia marina

177

140(K), Jef 356(S), Sch 11655 (K), Lam 403(K), Rob 2285,2548, 2678,2855,3111(S)

Localities recorded: M,S,P,F,N,C,An,Coe,Alp,Ast

Erect woody herb to 1 m with opposite ovate leaves with sharply serrated teeth and sharp point to 9 cm long and usually dark green, with long slender spike to 3 mm wide with dark blue flowers, a few opening at a time.

Tectona L.f.

T. grandis L.f.   'teak'

Bailey

Jef 744(S)

Localities recorded: M

Cultivated timber tree with ovate leaves to 50 cm long and large loose sprays of white flowers, each to 8 mm wide, with globose woody fruit enclosed in a bladder-like covering ripening from pale green to brown.

Vitex L.

V. doniana Sweet (*V. cuneata*)   'bois mozambique'

Summerhayes

Rob/Kea 3160(S), Pro 4453(K)

Localities recorded: M

Indigenous tree to 20 m with rounded crown, 5-foliolate leaves, the leaflets obovate to 15 cm long, and axillary clusters of small mauve and green flowers with oblong edible black fruit to 2 cm long.

V. trifolia L.   'bois noux noux'

Summerhayes, Wilson (2), Robertson (4)

Pro 4052(S),4452(K), Rob 2351(S)

Localities recorded: M,S,F,C,Fe

Shrub to 3 m with opposite trifoliolate leaves, the largest central leaflet to 8 cm long and the leaflets dark green above and pale greyish green below, sprays of pale blue flowers with greyish calyces, each to 1 cm long, and globose black fruit to 5 mm wide held in the enlarged grey calyx. (Fig. 135)

AVICENNIACEAE (125/5)

Avicennia L.

A. marina (Forsk.)Vierh.   'manglier blanc'

Summerhayes, Bailey, Piggott, Gwynne, Lionnet (1), Renvoize, Fosberg (2)

Jef 494,1232(S), Fos 52048(S), Rob 2451(S)

Localities recorded: M,C,Co,Ald,Far,Cos,Ast

Indigenous small tree, a mangrove, with pencil-like aerial roots rising out of the mud, opposite elliptic leaves, pale yellowish green to 10 cm long, clusters of yellow star-shaped fragrant flowers each to 8 mm wide and greyish green oval flattened fruits to 2.5 cm long which shed the outer grey skin as soon as they fall. (Fig.136)

NESOGENACEAE (125/6)

Nesogenes A.DC.

N. prostrata (Benth.)Hemsley (*N. dupontii*)

Fosberg (2)

Localities recorded: Ald,Ass,Cos,Ast

Endemic hairy herb with opposite or subopposite narrow ovate leaves to 2 cm long, solitary axillary mauve funnel-shaped flowers to 1 cm long and ovoid compressed hard fruit to 2 mm wide held in the persistent calyx.

LABIATAE (*LAMIACEAE*) (126)

Achyrospermum Blum.

A. seychellarum Baker

Summerhayes, Procter (1)

Jef 1149(S)

Localities recorded: M(? no longer),S

Endemic stout herb with 4 angled stems, opposite serrated edged ovate leaves to 20 cm long with long petioles to 7 cm, dense axillary cylindrical spikes to 4 cm long of tubular red flowers to 1 cm long and the fruit of 4 nutlets contained in the persistent calyx.

Coleus Lour. (probably to be included in *Plectranthus*)

C. amboinicus Lour. (*Plectranthus aromaticus*)   'baume, indian borage'

Bailey

No locality recorded

Cultivated small hairy succulent shrub to 50cm with ovate serrated edged leaves and spikes of small purple flowers. Used as a pot herb.

C. prostratus Guerke

Rob 3161(S)

Localities recorded: M

Fleshy creeping herb, often grown as an ornamental, with small succulent serrated edged leaves to 1 cm long and short spikes to 3 cm long of whorls of small pale mauve two-lipped flowers.

C. subfrutectosus Summerhayes (*C. suffruticosus* is a misspelling)   'gros baume'

Summerhayes, Procter (1)

Jef 652(S), Rob 2591(S), (? also Fos 52057(S), Bru/War 77(S), Rob 2675(S))

Localities recorded: M(? no longer),S(? no longer),N,?F,?A,?Co

Endemic fleshy creeping stout hairy herb with ovate tooth edged leaves to 10 cm long and long spikes of whorls of pale mauve flowers. (The taxonomy of *Coleus* is difficult and these specimens seem very close to *C. amboinicus*)

Leonotis (Pers.)R.Br.

L. nepetifolia R.Br.   'dacca, monte au ciel, lion's ear'

Summerhayes, Bailey, Lionnet (1), Renvoize, Fosberg (2), Baumer, Robertson (5),(6)

Rob 3130(S)

Localities recorded: P,N,An,Coe,Ald,Ass

Robust annual herb to 3 m with square stems, ovate or triangular serrated edged leaves to 11 cm long with the petioles to 10 cm long, terminal spikes bearing dense globose clusters, up to 6 cm wide, of bright orange hairy flowers and stiff spiny bracts, the 4 nutlets held in the persistent calyx. Used medicinally in Seychelles. (Fig. 137)

Leucas Burm.

L. aspera Link   'herbe madame tombé'

Summerhayes, Bailey, Evans

Localities recorded: M,P

Hairy herb with square stem and opposite leaves and axillary whorls of hairy white flowers, each to 1 cm long, the calyx with stiff hairs.

L. lavandulifolia Sm.   'herbe madame tombé'

Summerhayes

Jef 654(S), Pro 4154(S), Rob 2601,2652(S)

Localities recorded: M,S,F,N

Hairy herb with lax or erect square stems, narrowly lanceolate opposite leaves to 10 cm long and axillary whorls of hairy white flowers, each to 1 cm long, the calyx sharply toothed but without stiff hairs. (Fig. 138)

Mentha L.

M. arvensis L.   'menthe, japanese peppermint'

Bailey

Ves sn(K), Dup 19/5/37(K)

Localities recorded: M(BG)

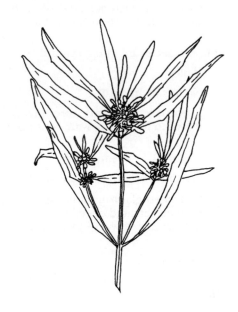

Fig. 137. Leonotis nepetifolia

Fig. 138. Leucas lavandulifolia

Fig. 139. Pogostemon cablin

Fig. 140. Boerhavia coccinea

Prostrate glabrous strongly aromatic herb with stems to 40 cm, the tips ascending, ovate serrated edged leaves to 4 cm long and rarely flowering, the flowers in whorls at the leaf axils and each to 3 mm long.

M. × piperita L.   'peppermint'

Bailey

No locality recorded

Erect perennial herb, strongly aromatic, with stalked ovate serrated edged leaves to 6 cm long with dense oblong terminal spikes of lilac pink flowers.

M. spicata L.   'spearmint'

Localities recorded: M(sr only)

Cultivated glabrous stoloniferous perennial herb with red stems and ovate pointed serrated edged sessile leaves to 5 cm long, rarely flowering but with pale mauve flowers in whorls on long terminal spikes. Used as a herb.

Ocimum L.

O. basilicum L.   'basilic de france, toc maria, sweet basil'

Summerhayes, Bailey, Fosberg,(2), Robertson (2),(4)

Localities recorded: M,F,Coe,Ald

Cultivated glabrous herb with square stems, ovate slightly serrated opposite leaves to 8 cm long, very pungent, and spikes of white flowers, each to 8 mm long, the calyx not inflated. Used as a herb.

O. canum Sims   'basilic'

Fosberg (2)

Dup 1/20,2/20(K) (named as *O. americanum*)

Localities recorded: ?M,Ast

Cultivated herb similar to *O. basilicum* but with flowers not exceeding the calyx, about 4 mm long. Used as a herb.

O. gratissimum L.   'gros basilic, basilic grande feuille'

Summerhayes, Bailey, Renvoize, Fosberg (2)

Pro 4449(K), Rob 2612(S)

Localities recorded: M,S,N,Ald

Cultivated aromatic perennial woody herb to 1 m with ovate pointed serrated edged opposite leaves to 10 cm long, the petioles to 5 cm long, long slender spikes of small white flowers and the calyx inflated around the 4 nutlets.

O. sanctum L.   'basilic petite feuille, sacred basil'

Summerhayes, Bailey, Fosberg (2)

Dup 16/34(K), ?Dup 1937(K)

Localities recorded: M,S,Ald

Cultivated herb similar to *O.basilicum* but with stiffer hairs on the stem.

O. viride Willd.

Dup 4/17/(K)

Localities recorded: ?M

Cultivated herb similar to *O. gratissimum* but with midrib of leaf hairy beneath.

Pogostemon Desf.

P. cablin (Blanco)Benth.   'patchouli'

Bailey, Lionnet (1)

Localities recorded: M(sr only)

Cultivated and escaping woody pubescent herb to 1 m, with rhomboid coarsely tooth edged leaves to 11 cm long and dense spikes of pinkish purple flowers each to 8 mm long in whorls. An essential oil is distilled from the leaves which can also be used dry in sachets. (Fig. 139)

Salvia L.

S. coccinea J.Murr.   'salvia'

Bailey

No locality recorded

Ornamental herb with opposite ovate stalked leaves to 5 cm long, spikes of scarlet two-lipped flowers, each to 2.5 cm long, in whorls and fruit of 4 3-angled ovoid nutlets.

S. hispanica L.   'blue salvia'

Dup 6/32(K)

Localities recorded: ?M

Ornamental herb similar to *S.coccinea* but with pale blue flowers in denser spikes.

Solenostemon Thonn.

S. scutellarioides (L.)Codd (*Coleus atropurpureus, Coleus blumei, Coleus scutellarioides*)   'amarante, coleus, painted nettle'

Summerhayes, Bailey

Jef 673(S)

Localities recorded: M

Ornamental herb with square stems, the young shoots succulent, pubescent pointed ovate tooth edged opposite leaves to 10 cm long, variously coloured, purple, red, yellow and green, and long spikes of whorls of blue flowers with a long lower lip, each flower to 1 cm long.

Thymus L.

T. vulgaris L.   'thym, thyme'

Bailey

Localities recorded: M(sr only)

Cultivated low woody perennial herb to 20 cm tall with minute elliptic leaves to 7 mm long and spikes of whorls of small blue flowers, each to 4 mm long. Used as a herb and yields thymol.

PLANTAGINACEAE (127)

Plantago L.

P. major L.   'plantain'

Pro 4380(S)

Localities recorded: S

Herb with basal rosette of ovate long-petioled leaves, the blade to 20 cm long, and long-stemmed cylindrical spikes of small sessile greenish white flowers, each to 4 mm long, with globose pale brown capsules to 3 mm long.

NYCTAGINACEAE (128)

Boerhavia L.

B. coccinea Mill.   'patate cauvii'

Gwynne, Renvoize

Jef 1181,1206(S), Bru/War 40(S), Ves 5401,5654(K)

Localities recorded: F,A,R,Bir,Den,Coe,Afr,Rem,Des,Pro,Far,Ald,Ass,Cos,Ast

Prostrate or scambling glandular pubescent herb, with ovate leaves to 5 cm long, small whitish flowers in long-stemmed clusters, the axillary stems with more than one cluster, and sticky fruits to 3 mm long. Closely related to *B. repens*. (Fig. 140)

B. crispifolia Fosberg

Renvoize, Fosberg (2)

Localities recorded: Ald

Endemic spreading herb from a thickened rootstock with ovate leaves to 1 cm long, clusters of 3 to 6 small pink flowers, each to 2.5 mm long, and 5-ribbed fruits with few or no glands.

B. diffusa L.

Summerhayes, Gwynne, Stoddard (1),(3)

Rob 2492(S)

Localities recorded: B,Bir,Coe,Rem,Dar,Far

Spreading or prostrate herb with stems to 30 cm with long hairs at the nodes and on the petioles, ovate or rounded leaves to 5 cm long, pale beneath, small purple flowers in long stemmed clusters and glandular sticky fruit to 5 mm long.

B. repens L.

Fosberg (2), Stoddart (1),(2),(3), Wilson ?(3),?(4)

Fos 52077,52145(S), Pro 4043(S), Rob 2506,2869,3035,3061(S)

Localities recorded: Co,Bir,Den,Pla,Coe,Afr,Rem,Dar,Jos,?Mar,?Def,Alp, Ald,Cos,Ast

Prostrate finely pubescent herb with ovate leaves to 2 cm long, clusters of small pinkish flowers, up to 5 per cluster and one cluster per stem, and glandular sticky fruits to 3 mm long. Closely related to *B. coccinea*.

(Note — the taxonomy of *Boerhavia* is confused and many of the specimens quoted above may not be correctly identified. The descriptions for B. coccinea, B. diffusa and B. repens were taken from Agnew)

Bougainvillea Comm.

B. glabra Comm.   'villea, bougainvillea'

Bailey, Stoddart (3), Robertson (3),(4),(7)

Localities recorded: M(sr only),F,Bir,Dar,Poi

Ornamental large strongly branched woody liane with scattered small ovate softly hairy leaves to 5 cm long, recurving thorns which are reduced flower spikes and 3 parted clusters of small yellowish white tubular flowers, each with a large showy purple bract.

B. spectabilis Willd.   'villea, bougainvillea'

Bailey

No recorded locality

Similar to above with many different colour varieties.

Commicarpus Standley

C. plumbagineus (Cav.) Standley (*Boerhavia africana*)

Fosberg (2)

Localities recorded: Ass

Semi-scandent herb, woody at base, with ovate to cordate leaves to 3 cm long, clusters of white trumpet-shaped flowers each to 7 mm long, with conspicuous stamens and stigma, and a 10-ribbed fruit to 3.5 mm long.

Mirabilis L.

M. jalapa L.   'belle de nuit, four o'clock plant, marvel of peru'

Bailey, Gwynne, Lionnet, (1), Renvoize, Fosberg (2), Stoddart (1),(2),(3), Wilson (4)

Fos 52180(S), Bru/War 116(S), Rob 2957,3129(S)

Localities recorded: M(sr only),A,Co,Bir,Coe,Des,Dar,Mar,Far,Ald

Cultivated and escaping erect herb with triangular ovate leaves to 10 cm long, terminal clusters of showy red or yellow spreading tubular flowers, each to 3.5 cm wide, with 5 stamens, and a hard black ovoid wrinkled fruit to 1 cm long enclosed by the persistent perianth base. (Fig. 141)

Pisonia L.

P. aculeata L.

Fosberg (2)

Localities recorded: Ald

Climbing shrub to 10 m long with hooked spines on the stem, ovate leathery leaves to 12 cm long, greenish flowers to 4 mm long in dense ($\male$) or lax ($\female$) clusters, and narrow 5-ribbed glandular fruits to 12 mm long.

P. grandis R.Br. (*P. viscosa*)   'mapou, bois mapou'

Summerhayes, Bailey, Gwynne, Renvoize, Stoddart (2),(3), Fosberg (2)

Fos 52063(S), Bru/War 61(S), Rob 3176(S), Sch 11754(K)

Localities recorded: M,S,P,A,Co,Bir,Den,Dar,Jos,Pie,Ald,Ass,Cos,Ast

Indigenous tree or shrub to 10 m with pale soft trunk, ovate leaves to 30 cm long, clusters of small white funnel-shaped flowers, each to 5 mm long, and cylindrical strongly 5-ribbed glandular sticky fruit to 1 cm long which adhere to the feathers of birds. (Fig. 142)

AMARANTHACEAE (130)

Achyranthes L.

A. aspera L. (*A. indica*)   'herbe sergent, ?chirugiens, (?herbe surgeon)'

Summerhayes, Bailey, Gwynne, Renvoize, Fosberg (2), Stoddart (1),(2),(3), Wilson (4), Robertson (3)

Ves 6149(K), Jef 1210(S), Pig sn(K), Fos 52173(S), Bru/War 23(S), Rob 2624,2866,3021(S), Tod 44(S)

Localities recorded: P,F,N,A,Co,Bir,Den,Pla,Coe,Afr,Rem,Dar,Jos,Poi, Mar,Alp,Pie,Far,Ald,Ass,Cos,Ast

Erect pubescent herb with opposite ovate leaves to 10 cm long, terminal slender spikes of small pinkish flowers surrounded by stiff shiny bracts and strongly reflexed pointed fruit to 6 mm long which detach easily and cling to clothing. (Fig. 143)

Fig. 141. Mirabilis jalapa

Fig. 142. Pisonia grandis

Fig. 143. Achyranthes aspera

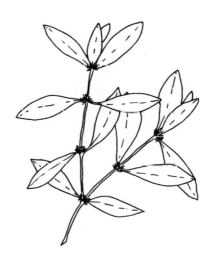

Fig. 144. Alternanthera sessilis

Aerva Forsk.

A. lanata (L.)Wall.

Summerhayes

Localities recorded: S

Erect or trailing pubescent herb with opposite and alternate rounded ovate leaves to 2 cm long and many short rounded axillary spikes of very small flowers and white bracts.

Alternanthera Forssk.

A. bettzickiana (Regel)Voss (*Telanthera bettzickeana*)

Summerhayes

Localities recorded: M

Ornamental border plant, a slender herb with opposite oblanceolate leaves to 5 cm long, variegated pink and bronzy, red stems and axillary clusters of flowers with silvery white bracts to 3 mm long.

A. pungens Kunth   'brede emballage'

Fosberg (2)

Localities recorded: Ald

Slender prostrate herb with opposite ovate leaves to 1.5 cm long, one larger than the other, and axillary clusters of small flowers with prickly straw coloured bracts, each to 6 mm long.

A. sessilis (L.)R.Br.   'brede emballage'

Summerhayes, Bailey, Gwynne, Renvoize, Wilson (1)

Jef 715(S), Pro 4158,4365(S), Bru/War 36(S), Rob 2467,3095(S)

Localities recorded: M,S,P,F,A,Coe

Slender prostrate herb often on wet ground, with opposite ovate to elliptic leaves to 6 cm long, sometimes red stems, and axillary clusters of flowers with silvery white bracts, each to 2 mm long. (Fig. 144)

Amaranthus L.

A. dubius Mart. ex Thell. (*A. caudatus*, *A. oleraceus*)   'brede malabar, brede parietare'

Summerhayes, Bailey, Gwynne, Renvoize, Fosberg (2), Stoddart (1),(3), Wilson (3),(4), Robertson (7)

Ves 5398(K), Jef 655(S), Fos 52034(S), Pro 4221(S), Fra 449(S), Bru/War 43(S), Rob 2306,2363,2261,3012,3096(S), Tod 15(S)

Localities recorded: M,F,N,A,Co,R,Bir,Den,Pla,Coe,Rem,Dar,Poi,Mar,Def, Pro,Far,Ald,Ass,Cos,Ast

Stout branched erect herb to 2 m with alternate ovate long-petioled leaves, the lower ones to 25 cm long, and branched sprays of densely flowered spikes producing shiny small black seeds from small capsules which split around the middle. The leaves are eaten as spinach.

A. viridis L.   'brede malabar'

Renvoize, Fosberg (2)

Pro 4221(K)

Localities recorded: Def,Ald,Ass,Cos

Erect glabrous annual to 60 cm with ovate leaves with a notched tip and sprays of densely flowered spikes producing small dark brown seeds from indehiscent capsules. The leaves are eaten as spinach.

Celosia L.

C. argentea L.   'crete de coq, cock's comb'

Bailey

Localities recorded: M(sr only)

Erect glabrous annual ornamental herb to 1.5 m with lanceolate leaves to 14 cm long and oblong or globose spikes of densely packed flowers with pale pink or white shining bracts. *C. cristata* is the cultivated 'cock's comb' with variously fasciated spikes in yellow, pink, red or purple, and is also grown on Mahe.

Cyathula Lour.

C. prostrata (L.)Blume   'sergent rouge'

Summerhayes

Jef 477(S), Rob 2689(S)

Localities recorded: M,S,F

Annual branched herb, prostrate to erect, with opposite ovate leaves to 8 cm long and slender spikes of small greenish flowers, cylindrically arranged at the tips and in clusters lower down, reflexed in fruit, each flower with pointed bracts to 3 mm long.

Deeringia R.Br.

D. polysperma (Roxb.)Moq.

Renvoize, Fosberg (2)

Localities recorded: Ald

Straggling shrub to 3 m with soft wooded stem, ovate alternate leaves to 3 cm long and flowers in loose axillary spikes, pale yellowish green and to 3 mm long, with white fleshy globose fruit to 5 mm wide containing many small black glossy seeds.

Gomphrena L

G. globosa L.   'immortelle'

Bailey

No locality recorded

Cultivated erect pubescent annual herb with opposite leaves to 11 cm long and globose heads to 2.5 cm wide of densely crowded white, pink or purple flowers and bracts.

Iresine P.Br.

I. herbstii Hook.f.   'iresine'

Bailey

Localities recorded: M(sr only)

Ornamental woody herb to 2 m with dark red bowl shaped leaves with bright red veins, or light green leaves with yellow veins, and yellowish white minute flowers and bracts.

Lagrezia Moq.

L. oligomeroides (C.H.Wright)Fosberg

Renvoize, Fosberg (2)

Fos 52075(S), Pro 4442(K)

Localities recorded: Co,Ald,Ass,Cos

Erect or spreading herb to 15 cm tall with alternate leaves to 2 cm long (longer in Cousin specimens), slightly fleshy, with clusters of small pinkish flowers and indehiscent capsules, each with one small shiny black seed.

CHENOPODIACEAE (131/1)

Arthrocnemum Moq.

A. pachystachyum (Bunge ex Ung.-Sternb.)A.Chew.

Fosberg (2)

Localities recorded: Ald

Much branched small leafless glaucous shrub with swollen jointed green stems and dense spikes bearing normal and ♂ flowers and vertically compressed seeds.

Beta L.

B. vulgaris L.   'betterave, beetroot'

Bailey

Localities recorded: M(sr only)

Cultivated vegetable with swollen globose edible red root and large green leaves with red midribs in a rosette.

Chenopodium L.

C. ambrosioides L. (*C.anthelminticum*) 'ambrosine, botrys, semen contrat, wormseed'

Summerhayes, Bailey

Localities recorded: M

Erect branched herb to 1 m with an aromatic odour, angled stems, wavy edged lanceolate leaves to 10 cm long and long leafy spikes of tight clusters of small sessile greenish flowers and capsules containing compressed seeds to 1.2 mm wide.

BASELLACEAE (131/2)

Basella L.

B. alba L. (*B. rubra*)   'brede gandolle, indian spinach'

Bailey

Localities recorded: M(sr only)

Cultivated perennial twining herb with fleshy wide ovate or rounded leaves to 20 cm long, axillary spikes of small whitish flowers and black fleshy flattened globose fruit to 1 cm wide. There are green and red tinted varieties and the leaves are eaten as spinach.

POLYGONACEAE (134)

Antigonon Endl.

A. leptopus Hook. & Arn.   'antigone, coral vine or creeper, bride's tears'

Bailey, Lionnet (2)

Jef 607(S)

Localities recorded: M

Cultivated ornamental climber with tuberous roots, triangular leaves to 12 cm long, hanging sprays of pink or white flowers, each to 1 cm long with two large and three small segments, and ovoid capsules to 1 cm long containing small brown seeds, the flower sprays ending in tendrils. (Fig. 145)

Coccoloba L.

C. uvifera L.   'sea grape'

Bailey

Jef 1159(S)

Localities recorded: M(BG)

Ornamental tree with leathery kidney-shaped leaves to 15 cm long with red veins and stipular sheaths, stiff spikes of spaced-out white flowers and clusters of globose violet fruit to 2 cm wide, astringent but edible.

Polygonum L.

P. senegalense Meisn.   'persicaire'

Summerhayes

Jef 659(S), Pro 4256(S), Rob 2805(S)

Localities recorded: M,S,P,D,N,Co

Indigenous erect herb of marshy areas with stems rooting at the nodes and lanceolate leaves to 25 cm long with red petioles and brown stipular sheaths, long stalked slender spikes of white or pink flowers, each to 3 mm long, and capsules with black glossy seeds. (Fig. 146)

NEPENTHACEAE (136)

Nepenthes L.

N. pervillei Blume   'liane pot a eau, pitcher plant'

Summerhayes, Bailey, Procter (1), Lionnet (1)

Ves 5516A(K), Arc 137(K), Jef 559(S),463(K), Sch 11731(K), Fos 51998(S), Schmid Hollinger 82-87,93,95, 97,98(S), Ber 14677(K), Rob 2478(S)

Localities recorded: M,S

Endemic woody climber with rosettes of leathery ovate leaves to 25 cm long, the midrib extended to form a tendril or a vertically held funnel shaped pitcher to 15 cm tall with a lid. The brownish cream ♂ and ♀ flowers are in loose clusters on erect separate stalks to 25 cm tall and the fruit are club-shaped capsules to 1.5 cm long. (Fig. 147)

ARISTOLOCHIACEAE (138)

Aristolochia L.

A. elegans Mast.   'dutchman's pipe'

Localities recorded: M(sr only)

Ornamental climber with triangular to heart-shaped leaves to 10 cm long, flowers with an inflated bent tube and wide heart-shaped limb, mottled brown and purple and to 8 cm wide, and oblong ridged capsules which open like a basket and containing flat brown seeds.

A. gigantea Mart.   'giant dutchman's pipe'

Localities recorded: M(sr only)

Ornamental climber similar to *A.elegans* but with flowers to 20 cm or more wide.

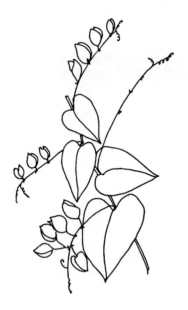

Fig. 145. Antigonon leptopus

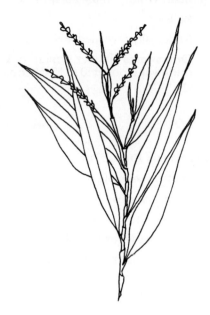

Fig. 146. Polygonum senegalense

Fig. 147. Nepenthes pervillei

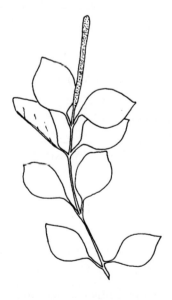

Fig. 148. Peperomia dindigulensis

A. hians Willd.   'liane cochon'

Bailey

No locality recorded

Ornamental climber similar to *A. elegans* but flowers with a two-lipped limb, the lower elongated to 15 cm long and the upper with two rounded lobes to 7 cm wide.

PIPERACEAE (139/1)

Peperomia Ruiz & Pav.

P. dindigulensis Miq.

Summerhayes

Jef 504(K)

Localities recorded: M

Indigenous erect succulent pubescent herb with obovate pointed leaves to 2 cm long and thin spikes to 5 cm long, of tightly packed minute greenish flowers. (Fig. 148)

P. pellucida (L.)H.B.K.

Jef 1174(S), Bru/War 34(K), Tod 51(S)

Localities recorded: M(sr only),F,A

Ornamental and escaping annual succulent herb with pale green juicy stems, ovate leaves, almost translucent to 3 cm long, and erect green spikes to 6 cm long of tightly packed minute flowers and globose fruits less than 1 mm long.

P. portulacoides A.Dietr.

Summerhayes

Localities recorded: M

Indigenous erect succulent herb similar to *P. dindigulensis* but with leaves to 4 cm long and flower spikes to 10 cm long.

Piper L.

P. betle L.   'betel'

Summerhayes, Bailey, Baumer

No locality recorded

Climber with swollen nodes and heart-shaped leaves to 15 cm long with sheathing petioles, ♂ flowers in spikes to 10 cm long and 2 mm wide, ♀ flowers red and fleshy in spikes to 5 cm long and 1cm wide. The leaves are chewed with betel nut as a masticatory. Used medicinally in Seychelles.

P. nigrum L.   'poivre, black pepper'

Summerhayes, Bailey

Localities recorded: M(sr only)

Climber with heart-shaped leaves to 18 cm long with sheathing petioles, minute flowers, usually ♂ and ♀ together on spikes and red globose fruit, each to 5 mm wide. Black pepper is ground from the whole dried fruit, the skin is removed before grinding white pepper. Unripe fruits are also used as a flavouring.

P. radicans Vahl   'liane de poivre, poivre marron'

Bailey

No locality recorded

Climber with heart-shaped to pointed ovate leaves to 5 cm long and pubescent on petiole and veins beneath.

P. umbellatum L.

Summerhayes

Sch 11669(K), Pro 4091(S)

Localities recorded: M

Indigenous semi-climbing shrub to 3 m with large heart-shaped leaves to 30 cm wide and clusters of narrow cylindrical spikes to 8 cm long of densely arranged minute white flowers.

MYRISTICACEAE (141)

Myristica Gronov.

M. fragrans Houtt.   'muscadier, nutmeg'

Bailey, Lionnet (2)

Localities recorded: M(BG)

Cultivated tree with rounded crown to 20 m with shiny ovate leaves to 10 cm long and small yellow bell-shaped flowers to 1 cm long, ♂ flowers on one tree, ♀ on another, with globose yellow fruit to 7 cm long containing one seed (the nutmeg) surrounded by a red aril (the mace), both of which are spices.

LAURACEAE (143/1)

Cassytha L.

C. filiformis L.   'liane sans fin, dodder'

Summerhayes, Bailey, Gwynne, Renvoize, Fosberg (2), Stoddart (1),(3), Wilson (2), Robertson (3)

Jef 781(K), Sch 11686(K), Rob 2682,2873,2952,3006,3064(S)

Localities recorded: M,P,F,C,L,Bir,Den,Pla,Coe,Afr,Rem,Des,Dar,Jos, Poi,Alp,Pro,Far,Ald,Ass,Cos,Ast

Indigenous leafless plant with string-like orange, yellow or green stems, clinging to other vegetation with suckers, with scattered small white flowers on short spikes and white globose fruit to 5 mm wide with the remains of the calyx at the apex.

Cinnamomum Schaeffer

C. camphora (L.)Nees & Eberm.   'camphor'

Summerhayes, Bailey

Localities recorded: S

Cultivated tree to 30 m with ovate leaves, whitish beneath, with one main vein, to 10 cm long, cone-like leafy buds, small pale yellow flowers to 4 mm wide and black fleshy globose fruits to 1 cm wide held in the smooth conical calyx. Camphor oil is distilled from the wood.

C. zeylanicum Blume   'cinnamon, cannelier'

Summerhayes, Bailey, Sumner Gibson, Lionnet, (1), Wilson (2), Robertson (6)

Rob 2255,2667(S)

Localities recorded: M,S,P,D,F,C,An

Cultivated and escaping shrub or tree to 17 m with aromatic bark and leaves, leaves stiff and glossy with 3 main veins to 17 cm long, reddish when young, sprays of cream flowers, on cream stems, each to 3 mm wide, and dark blue or black ovoid fleshy fruit with one large seed, to 1.5 cm long, held in a cup-like calyx. The bark is powdered to form the spice and oils can be distilled from the leaves and from the bark. (Fig. 149)

Litsea Lam.

L. glutinosa (Lour.)C.B.Robinson   'bois d'oiseau, bois zoizo'

Summerhayes, Bailey, Robertson (6)

Jef 674(S), Rob 3184(S), Tod 21(S)

Localities recorded: M,F,An

Tree to 10 m with elliptic leaves to 20 cm long, few-flowered axillary clusters of yellowish, many-stamened flowers, each to 8 mm wide, in the upper axils, and globose fruit to 8 mm wide.

Persea Mill.

P. americana Mill.   'avocat, avocado, alligator pear'

Summerhayes, Bailey, Fosberg, Baumer, Robertson (4),(5),(7)

Localities recorded: M(sr only),F,N,Co,Dar

Cultivated tree to 20 m with ovate pointed leaves to 30 cm long, small clusters of yellow flowers, each to 1cm wide, and large green to purple pear-shaped

Fig. 149. Cinnamomum zeylanicum   Fig. 150. Hernandia nymphaeifolia

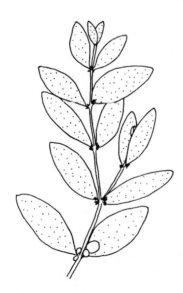

Fig. 151. Viscum triflorum   Fig. 152. Acalypha indica

197

globose fruit to 20 cm long with edible greenish yellow oily flesh and one large seed. Used medicinally in Seychelles.

HERNANDIACEAE (143/2)

Hernandia L.

H. nymphaeifolia (Presl)Kubitzki (*H. ovigera, H. peltata*)   'bois blanc'

Summerhayes, Bailey, (not *H.sonora*), Piggott, Gwynne (not *H.sonora*), Renvoize, Fosberg (2), Stoddart (not *H.sonora*) (1),(2),(3), Wilson (4), Robertson (3),(4), (6)

Jef 681(S), Bru/War 82(S), Rob 2300,2827,3034,3057(S),2383(K)

Localities recorded: M,P,F,A,An,Bir,Den,Pla,Coe,Rem,Des,Dar,Jos,Poi, Mar,Alp,Pro,Far,Ast

Indigenous coastal tree with pale bark, peltate leaves to 30 cm long, separate ♂ and ♀ trees, the flowers cream coloured to 3.5 mm long and with black globose fruit to 2.5 cm wide encased in an inflated cup to 5 cm long, usually whitish green. (Fig. 150)

PROTEACEAE (144)

Grevillea R.Br.

G. robusta A.Cunn.   'silky or silver oak'

Bailey, Swabey

Localities recorded: M

Cultivated shade or timber tree to 40 m with fern-like pinnate leaves with lobed leaflets with silky hairs and appearing greyish, and orange flowers in one-sided clusters.

THYMELAEACEAE (145/1)

Wikstroemia Endl.

W. indica (L.)C.A.Meyer (*W. viridiflora*)   'herbe tourterelle'

Bailey

No locality recorded

Cultivated shrub to 3 m with ovate leaves to 5 cm long with prominent veining and few-flowered terminal clusters of softly hairy greenish white flowers to 1.2 cm long and ovoid reddish fruit to 6 mm long.

LORANTHACEAE (including *VISCACEAE*) (148)

Bakerella van Tiegh.

B. clavata (Desr.)S.Balle ssp. **clavata** (*Loranthus clavatus*)   'mangoula'

Renvoize, Fosberg (2)

Localities recorded: Ald

Parasitic shrub with ovate to sickle-shaped leaves to 9 cm long, with narrow tubular flowers to 3 cm long, the tube red and the lobes yellow, and ovoid green fruits to 1.8 cm long, containing one seed with star-like apex.

B. clavata (Desr.)S.Balle ssp. **sechellensis** (Baker) S.Balle (*Loranthus sechellensis*) 'parasite, bois marmaille'

Summerhayes, Bailey, Procter (1)

Localities recorded: M,S,P,C

Endemic (possibly extinct) parasitic shrub similar to ssp. *clavata* but with ovate leaves to 2 cm long and flowers to 4 cm long.

Korthalsella van Tiegh.

K. opuntia (Thunb.)Merr. (*Viscum opuntia*)

Summerhayes

Localities recorded: M

Indigenous parasitic plant with flattened and jointed green stem-sections, each about 1 cm long.

Viscum L.

V. triflorum DC.   'bois mamai, gui du pays'

Summerhayes, Renvoize, Fosberg (2)

Procter 4086,3990(S), Rob/ChongSeng 2822A(S)

Localities recorded: M,P,Ald

Indigenous parasitic shrub with swollen nodes, elliptical leathery leaves to 6 cm long, flowers in axillary pairs to 3 mm long, and pale green to white ovoid or globose fruit to 5 mm long containing one seed. (Fig. 151)

EUPHORBIACEAE (151/1)

Acalypha L.

A. claoxyloides Hutch.   'bois malgache'

Renvoize, Fosberg (2)

Localities recorded: Ald,Ass,Cos,Ast

Endemic irregularly branched shrub with resinous glands, bright green ovate variable leaves to 15 cm long, ♂ and ♀ flowers on separate spikes, often on different plants, ♂ spikes with scale-like bracts and ♀ spikes with 1 to 5 large leafy bracts.

A. hispida Burm.f. (*A. sanderiana*)   'queue de chatte, queue de mimi, red hot cat's tails'

Bailey, Lionnet (2), Robertson (7)

Ornamental shrub to 3 m with heart-shaped leaves to 20cm long and long drooping furry tail-like spikes of red ♀ flowers with multi-branched stigmas.

A. indica L. 'herbe chatte'

Summerhayes, Bailey, Gwynne, Renvoize, Fosberg (2), Stoddart (1),(3), Baumer, Wilson (2),(3),(4), Robertson (7)

Jef 709(S), Lam 408(K), Pro 4045,4224(S), Fos 52070(S), Bru/War 35(S), Rob 2380,2443,2716,2853,3044,3146(S)

Localities recorded: M,F,C,A,Co,Bir,Den,Pla,Coe,Rem,Dar,Poi,Mar,Def, Alp,Far,Ald,Ass,Cos,Ast

Erect annual herb, a weed, to 50 cm tall with rhomboid long-stalked leaves to 10 cm long with axillary spikes of flowers with the ♂ flowers small and at the tip and the ♀ flowers with large leafy bracts, each to 7 mm long, down the rest of the spike. Used medicinally in Seychelles. (Fig. 152)

A. integrifolia Willd. (*A. colorata*)   'bois queues de rats'

Bailey

No locality recorded

Shrub to 2 m with oblong leaves to 15 cm long, the ♂ flowers in dense clusters on axillary spikes and ♀ flowers in bracts to 3 mm wide in the leaf axils. (Bailey's use of this name may be incorrect).

A. wilkesiana Muell.-Arg. (?*A. grandis*, *A. tricolor*)

Summerhayes, Bailey, Robertson (7)

Jef 606(S)

Localities recorded: M,Dar

Ornamental shrub to 4 m with variable leaves with many colour variations and shapes, but usually ovate and serrated edged, with ♂ and ♀ spikes on the same plant, the ♂ flowers small with minute bracts and the ♀ flowers with pink stigmas and leafy bracts to 5 mm long.

Aleurites J.R. & G.Forst.

A. moluccana (L.)Willd. (*A. triloba*)   'bankul, candle nut tree'

Bailey, Baumer

Rob 2802(S)

Localities recorded: M(sr only),P

Cultivated and escaping tree with 5-lobed leaves to 20 cm long, greyish beneath, terminal clusters of ♂ and ♀ flowers, creamy white to 6 mm long, and fleshy ovoid fruit to 6 cm long containing one or two hard shelled oily seeds which yield a commercial drying oil. Used medicinally in Seychelles.

200

Breynia J.R. & G.Forst.

B. disticha Forst. var. **disticha** f. **nivosa** (Bull) Croizat (*B. nivosus* var. *roseo-pictus*) 'boule de neige, snow bush'

Bailey

Dup 70/07(K)

Localities recorded: M(sr only)

Ornamental shrub to 2 m with variegated green and white, often pinkish, rounded to ovate leaves to 5 cm long and small green flowers. Often used as a hedge plant.

Codiaeum A.Juss.

C. variegatum (L.)Blume 'croton'

Summerhayes, Bailey, Wilson (1), Robertson (3),(7)

Jef sn(K)

Localities recorded: M,S,P,Dar,Poi

Ornamental shrub to 5 m with glossy colourful leaves, green to red to yellow, spotted or variegated and variable in shape, with slender axillary spikes to 25 cm long of small yellowish flowers, ♂ and ♀ separate.

Drypetes Vahl

D. riseleyi Airy Shaw (*Riseleya griffithii*) 'bois marais petite feuille'

Summerhayes, Bailey, Procter (1)

Ves 5533,6132(K), Jef 696,1148(K), Pro 3983,3984(S)

Localities recorded: M,S,P

Endemic tree to 15 m with ovate to oblong leathery leaves to 20 cm long, shortly stalked axillary pubescent flowers to 1.5 cm wide and globose golden fruit, pubescent, to 5 cm long and containing 2 large seeds.

Euphorbia L.

E. cyathophora Murr. (*E. heterophylla*) 'wild poinsettia'

Summerhayes, Gwynne, Stoddart (1), (3)

Jef 1204(S), Sch 11775(K), Pro 4145(S), Rob 2356,2505(S)

Localities recorded: M,S,P,Bir,Den,Des,Dar,Poi

Annual herb, a weed to 1 m with variable leaves, the upper ones fiddle-shaped, the lower ovate, to 15 cm long, terminal clusters of yellow flowers and the leaves and bracts around the flowers bright red. (Fig. 153)

E. hirta L. 'jean robert'

Summerhayes, Bailey, Gwynne, Renvoize, Fosberg (2), Stoddart (1),(2),(3), Baumer, Wilson (1),(2),(4), Robertson (6)

Jef 706(S), Fra 423(S), Fos 52122(S), Bru/War 31(S), Rob 2246,2382, 2559,2641,2826,3084,3048(S)

Localities recorded: M,S,P,F,N,C,A,An,L,Co,Bir,Den,Pla,Coe,Rem,Des,Dar Jos,Poi,Mar,Alp,Pro,Far,Ald, Ass,Cos,Ast

Prostrate, spreading or erect herb, a weed, with milky juice, unequally based serrated edged leaves to 2.5 cm long in two rows, and axillary clusters of small flowers, the clusters flat and to 1 cm wide. Used medicinally in Seychelles. (Fig. 154)

E. inaequilatera Sond.

Rob 2524 (S)

Localities recorded: M

Prostrate glabrous herb with ovate serrated edged leaves to 6 mm long with long stipules and minute flowers with conspicuous white bracts.

E. indica Lam.

Fosberg (2)

Localities recorded: Ass

Erect or rarely spreading herb with oblong leaves to 1.5 cm long and loose clusters of small flowers, each cluster with two small leaves.

E. lactea Haw.

Localities recorded: M(sr only)

Large cactus-like plant to 5 m with thick green stems, triangular in section, the angles bearing projections with two small black spines and a small ovate leaf (on the young branches only). Used as a hedge plant.

E. mertonii Fosberg

Fosberg (2)

Localities recorded: Ald

Endemic glabrous prostrate herb with ovate purplish leaves to 1cm long and small flowers appearing as if on stems from the nodes.

E. milii Des Moul. var. splendens (Boj. ex Hassk.)Ursch & Leandri 'couronne d'epines, christ's thorn'

Bailey

Localities recorded: M(sr only)

Ornamental fleshy scambling shrub with spiny stems, a few ovate leaves to 4 cm long at the tops, and small yellowish flowers backed by two brilliant red bracts to 1 cm wide.

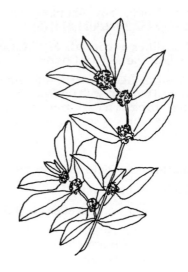

Fig. 153. Euphorbia cyathophora   Fig. 154. Euphorbia hirta

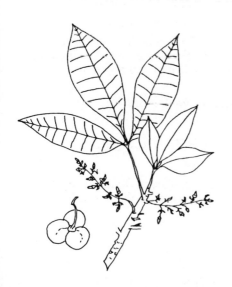

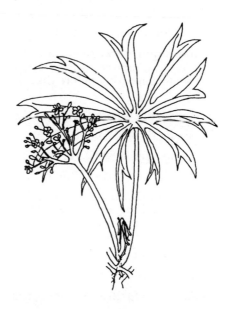

Fig. 155. Hevea brasiliensis   Fig. 156. Jatropha multifida

E. prostrata Ait.   'trainasse blanche, trainasse rouge'

Summerhayes, Bailey, Gwynne, Renvoize, Fosberg (2), Baumer, Stoddart (1),(3), Wilson (?1),(3),(4)

Jef 716(S), Fos 52148(S), Pro 4222,4230(S),4039(K), Fra 432(S), Rob 2376,2452,2862,3011,3147(S)

Localities recorded: M,?P,L,Co,Bir,Den,Pla,Coe,Rem,Des,Dar,Poi,Mar, Def,Alp,Far,Ald,Ass,Cos,Ast

Prostrate herb, a weed, with greenish or reddish tinge, the stems radiating from a tap root, with finely serrated edged ovate leaves to 5 mm long, the veins appearing as a dark green net, and minute flowers in small clusters in the nodes, the capsules hairy on the 3 angles only. Used medicinally in Seychelles.

E. pulcherrima Willd. ex Klotzsch   'poinsettia'

Bailey, Robertson (7)

Localities recorded: M(sr only),Dar

Ornamental shrub to 4 m with deeply cut leaves to 20 cm with the upper ones and the bracts around the flowers bright red, the small flowers red with two large yellow glands.

E. pyrifolia Lam.   'tonga, tanghin rouge, fangame, faudamon, bois de lait'

Summerhayes, Renvoize, Fosberg (2), Baumer, Wilson (1),(2)

Ves 5941,6120(K), Jef 585,626,628,691,1179(S), Fos 52156,52056,52046(S), Pro 3963(S), Fra 453,454(S), Ber 14651(K), Bru/War 62,64,81(S), Rob 2527,2658(S)

Localities recorded: M,S,P,F,C,A,Fe,Co,Cu,Pie,Ald,Ass,Cos,Ast

Indigenous deciduous glabrous shrub with irritant milky sap, elliptic leaves to 10cm long crowded at the branch tips, ♂ and ♀ flowers on separate plants and usually appearing before the leaves. Used medicinally in Seychelles.

E. stoddartii Fosberg (? close to *E.microphylla* Heyne)

Summerhayes, Fosberg (2)

?Coppinger 4/82(K), ?Fryer n/n(K), ?Ves 5963(K), ?Pro 4334(K)

Localities recorded: ?Den,?Pro,Ald,Ass,Cost,Ast

Endemic (?to Aldabra group) slender prostrate herb with slightly fleshy leaves to 3 mm long and solitary axillary minute flowers with apparently several stipules per node.

E. thymifolia L.

Summerhayes

Fos 52030(S), Bru/War 44(S), Rob 2615,2724(S)

Localities recorded: F,N,A,L,Co

Prostrate herb with reddish stems radiating from a taproot, ovate serrated edged pointed leaves to 8 mm long and flowers on short side shoots with smaller leaves and with hairy capsules.

E. tirucalli L.   'bois malgache, calli, tirucalli'

Bailey, Baumer

No locality recorded

Glabrous shrub or small tree with cylindrical green fleshy branches with irritant milky sap, a few small linear leaves and small flowers in the forks of the younger branches. Used as a boundary hedge and used medicinally in Seychelles

Excoecaria L.

E. benthamiana Hemsley   'bois charlot, bois jasmin rouge'

Summerhayes, Bailey, Lionnet (1), Procter (1)

Dup 13/32(K), Ves 5510(K), Jef 767(S),769(K), Sch 11832/3(K), Pro 3938,4264(S), Rob 2528(S)

Localities recorded: M,P

Endemic small tree with loose crinkly bark and leathery ovate leaves to 20 cm long clustered at the ends of the branches, very irritant milky sap, ♂ flowers in short spikes in the upper leaf axils with a ♀ flower at the base of each spike, the capsule 3-angled and to 1 cm long.

Hevea Aubl.

H. brasiliensis (Willd. ex A. Juss.)Muell.-Arg.   'caoutchouc, rubber'

Sumner Gibson, Bailey, Wilson (1), Robertson(4)

Rob 2426(S)

Localities recorded: M,S,P,F

Cultivated and escaping tree to 20 m with milky sap, long-petioled trifoliolate leaves, the leaflets elliptic to 20 cm long, and delicate sprays of cream flowers, the central flower in each cluster ♀, the others ♂ with a large woody 3-lobed capsule to 6 cm long containing hard mottled pale and dark brown shiny seeds to 2 cm long. (Fig. 155)

Jatropha L.

J. curcas L.   'medicinier, pignon d'inde, physic nut'

Summerhayes, Bailey, Swabey

Rob 2626,2711(S)

Localities recorded: M,P,F,N

Cultivated and escaping succulent shrub to 5 m with thick branches, 3–5-lobed leaves to 18 cm long, rounded clusters of greenish white flowers, mostly ♂, and yellow 3-lobed capsules containing 3 black seeds. The seeds yield a medicinal oil. Stakes were planted as vanilla supports and shade.

J. multifida L.   'arbre corail, coral plant'

Bailey, Robertson (7)

Localities recorded: M(sr only), Dar

Ornamental succulent shrub with large long-stalked deeply palmately 9–11-lobed leaves, dark green above, glaucous below, long stalked clusters of bright red, mostly ♂ flowers with 5 petals, and a 3-lobed capsule containing 3 seeds which splits explosively. (Fig. 156)

J. pandurifolia L.   'arbre corail'

Bailey

Jef 637(S)

Localities recorded: M

Ornamental shrub to 4 m with lax long-stalked ovate pointed leaves with a few teeth at the base, flowers in long stalked clusters, deep pink to 1 cm long, mostly ♂.

J. podagrica Hook.   'arbre corail petit, gout foot'

Bailey, Robertson (7)

Jef 598(S)

Localities recorded: M,Dar

Ornamental succulent subshrub, the stem with a swollen base like a bottle, with long-stalked peltate lobed leaves, glaucous below, and terminal clusters of bright red flowers, mostly ♂, and 3-lobed capsules to 1.5 cm long which explode to scatter the 3 ovoid seeds.

Mallotus Lour.

M.sp.

Summerhayes

Localities recorded: M

Shrub with obovate leaves to 8 cm long and a capsule to 8 mm long.

Manihot Mill.

M. esculenta Crantz (*M.utilissima*)   'manioc, cassava, tapioca'

Summerhayes, Bailey, Fosberg, Baumer, Wilson (2), Robertson (3),(4)

?Fryer 124/154(K), Bru/War 114(S)

Localities recorded: M(sr only),F,C,A,Co,Poi,?Ald

Cultivated shrub to 3 m with brittle knobbly stems from spreading tuberous roots, deeply palmately lobed leaves to 20 cm long, clusters of ♂ and ♀ flowers, ♂ to 8 mm long and ♀ to 1 cm, and globose narrowly winged capsules to 1.5 cm long containing 3 seeds. The tubers are an important source of starch. Used medicinally in Seychelles.

M. esculenta × **M. glaziovii** Muell.-Arg.   'tree cassava, ceara rubber'

Localities recorded: M(sr only)

Similar to *M. esculenta* but a tree without tuberous roots. There is much confusion over the naming of tree cassavas.

Margaritaria L.f.

M. anomala (Baill.)Fosberg (*Phyllanthus cheloniphorbe*)   'si là, bois doux'

Renvoize, Fosberg (2)

Localities recorded: Ald,Cos,Ast

Shrub or small tree with obovate leaves to 10 cm long, ♂ and ♀ plants, ♂ flowers in dense yellow axillary clusters, ♀ flowers on longer stalks in the leaf axils, and a fleshy globose fruit to 7 mm long containing 2 seeds in each of the 2 cells of the nut. Represented by an endemic variety on Aldabra, Cosmoledo and Astove.

Pedilanthus Neck. ex Poit.

P. tithymaloides (L.)Poit.   'bois mal gaz, slipper flower'

Gwynne, Renvoize, Fosberg (2), Stoddart (1),(3), Wilson (4), Robertson (3),(4),(7)

Bru/War 75(S), Rob 3051,3118(S)

Localities recorded: M(sr only),F,A,Bir,Den,Pla,Coe,Rem,Des,Dar,Poi, Mar,Far,Ald,Ass,Cos,Ast

Ornamental and escaping succulent subshrub with zigzag dark green branches and fleshy ovate leaves to 8 cm long with inconspicuous flowers inside bright pinkish red slipper shaped bracts. There is a green and white variegated form. It is grown near houses to ward off ill luck.

Phyllanthus L.

P. acidus (L.)Skeels (*P. distichus*)   'bilimbelle, cherimbolier, otaheite gooseberry'

Summerhayes, Bailey

Fos 52089(S), Bru/War 83(S)

Localities recorded: C,A,Co

Small tree with pointed ovate leaves to 9 cm long crowded on short branches which leave scars when they fall, small red flowers and greenish yellow waxy ribbed one-seeded fruit with acid edible flesh, borne on the leafless part of the branch.

P. amarus Schum. & Thonn. (*P. niruri*)   'curanellie blanche'

Summerhayes, Bailey, Fosberg (2), Stoddart (1),(3), Gwynne, ? Baumer, Wilson (2),(3),(4), Robertson (6)

Fos 52082(S), Pro 4136,4226(S), Fra 435,435A(S), Bru/War 29(S), Rob 2249,2345,2516B,2614,2713,2852, 3000,3083,3201(S)

Localities recorded: M,S,F,N,C,A,An,Co,Bir,Pla,Coe,Rem,Des,Dar,Poi,Mar,Def, Alp,Far,Ald,Ass,Cos,Ast

Erect slender herb with horizontal branchlets, small rounded to ovate almost sessile leaves to 1.5 cm long but usually less than 1 cm long, and shortly stalked capsules up to 2 mm wide with narrow sepals. ?Used medicinally in Seychelles. (Fig. 157)

P. casticum Willem. (*P. schimperianus*)   'bois castique, kirganellie'

Summerhayes, Bailey, Renvoize, Fosberg (2), Stoddart (1),(3), Wilson (1)

Fos 52031,52192(S), Bru/War 70(S)

Localities recorded: M,S,P,A,Co,Bir,Far,Ald,Ass

Spreading shrub or small tree with drooping arched branches, some leafy, some flowering, rounded ovate leaves to 5 cm long, small dark red ♂ and ♀ flowers in clusters on short branchlets, and shortly stalked globose red to black fruit to 4 mm wide.

P. maderaspatensis L. (incl. var. *frazieri* Fosberg)

Summerhayes, Gwynne, Renvoize, Fosberg (2), Stoddart (2),(3)

Jef 1198(S), Far 436(S), Rob 2297,2352,2837,3026,3087(S)

Localities recorded: P,Bir,Den,Pla,Coe,Rem,Des,Dar,Jos,Poi,Alp,Far,Ald,Ast

Slender herb with ascending branchlets, obovate to lanceolate leaves to 2 cm long or more, sometimes ending in a point, and capsules to 4 mm wide with wide sepals, to 1 mm or more wide. The endemic variety is on Aldabra and Astove. (Fig. 158)

P. mckenzei Fosberg   'canellie rouge'

Fosberg (2)

Localities recorded: Ald,Cos,Ast

Endemic delicate herb with slender branchlets, elliptic leaves to 5 mm long and the capsule to 1.5 mm wide with wide sepals.

P. niruroides Muell.-Arg.

Rob 3203(S)

Localities recorded: M

Slender herb, a weed, with stems angular in section, leaves obovate to 2 cm long and capsule to 2 mm wide with wide sepals with a broad stripe down each one.

P. nummulariifolius Poir. (*P. tenellus*)

Summerhayes, Stoddart (1)

Jef 501(S), Bru/War 28(K), War 187(S)

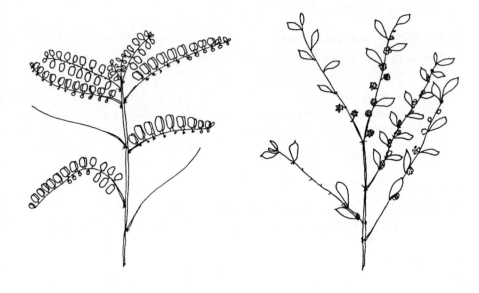

Fig. 157. Phyllanthus amarus   Fig. 158. Phyllanthus maderaspatensis

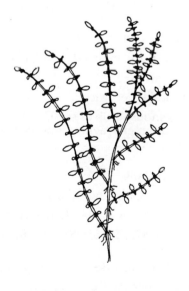

Fig. 159. Ricinus communis   Fig. 160. Pilea microphylla

Localities recorded: M,S,A,?Far

Slender herb with short erect branchlets, ovate leaves and small capsules, to 2 mm wide, on stalks to 6 mm long.

P. urinaria L.   'curanellie rouge'

Summerhayes, Bailey, Robertson (5)

Jef 711(S), Rob 2258,2714,3202(S)

Localities recorded: M,F,N

Slender herb, a weed, with stem sometimes narrowly winged, elongate oblong or ovate leaves to 2 cm long slightly curved forwards on horizontal branches which are red where they join the main stem, and capsules to 3 mm wide with narrow sepals and short stalks.

Ricinus L.

R. communis L.   'palma christi, ricin, tantan, castor'

Summerhayes, Bailey, Gwynne, Fosberg (1),(2), Stoddart (1),(2),(3), Wilson (2),(4), Robertson (6)

Rob 2835(S)

Localities recorded: C,An,Co,Bir,Den,Coe,Rem,Des,Dar,Mar,Alp,Far,Ald, Ass,Cos,Ast

Cultivated and escaping shrub to 5 m with palmately 6–11-lobed leaves to 50 cm long, terminal spikes of flowers with ♂ flowers below and ♀ flowers above, and a spiny capsule, globose, to 2.5 cm wide containing 3 mottled smooth seeds which yield castor oil. (Fig. 159)

Securinega Comm. ex Juss.

S. virosa (Roxb. ex Willd.)Baill. (*Flueggea virosa*)

Summerhayes

No locality recorded

Indigenous shrub or small tree to 6 m with elliptic leaves to 7 cm long, small creamy green fragrant flowers in axillary clusters and edible white fruit to 8 mm wide.

Wielandia Baill.

W. elegans Baill.   'bois fourmi'

Summerhayes, Bailey, Renvoize, Fosberg (2)

Ves 6199(K), Jef 835(S), Pro 4002,4245,4259,4377(S), 4552(K), Fos 52202(S), Ber 14639(K)

Localities recorded: M,S,P, (not Ald)

Indigenous small tree with leathery elliptic leaves to 8 cm long with petioles to 1 cm long, axillary clusters of white flowers on stalks to 3 cm long and hard 3-lobed capsules to 1 cm long.

URTICACEAE (153/1)

Boehmeria Jacq.

B. nivea (L.)Gaud.   'ramie'

Bailey

No locality recorded

Cultivated hairy subshrub with tuberous roots, ovate serrated edged pointed leaves, pale beneath, to 15 cm long, axillary sprays of flowers, each to 1.5 cm long, the lower ♂ and the upper ♀ and fruit a small nut to 1 mm long enclosed in the dry calyx. Cultivated for its fibre. Bailey may have confused this with *Pipturus argenteus*, or with *Laportea aestuans*.

Laportea Gaud.

L. aestuans (L.)Chew

Gwynne, Renvoize, Fosberg (2), Stoddart (1),(2), Wilson (4), Robertson (3)

Fra 426(S), Rob 2874,3040(S)

Localities recorded: Pla,Rem,Dar,Poi,Mar,Alp,Pro,Pie,Far,Ald

Hairy herb to 1 m with triangular serrated edged leaves to 15 cm long, spreading axillary sprays of yellowish green ♂ and ♀ flowers to 1.5 cm long and fruit to 2 mm long enclosed in the dry calyx.

Obetia Gaud.

O. ficifolia Gaud.   'figue marron'

Renvoize, Fosberg (2)

Localities recorded: Ald

Spreading tree to 6 m with ovate lobed serrated edged leaves to 20 cm long, usually clustered at the ends of the branches, clusters of pink ♂ flowers to 3 mm wide and ♀ yellowish flowers to 2 mm wide and ovoid fruit to 1 mm long enclosed in swollen perianth segments.

Pilea Lindl.

P. microphylla (L.)Leibm.   'artillery or gunpowder plant'

Rob 2423, 2861,3121(S)

Localities recorded: Coe,Poi,Alp

Succulent herb, often grown as an ornamental pot plant, but found on paths on coral islands, usually prostrate, with angular stems, many 2 ranked ovate leaves, each to 5 mm long, and tight axillary clusters of minute flowers, each to 1 mm wide, red to green, with explosive pollen. (Fig. 160)

Pipturus Wedd.

P. argenteus (Forst.)Wedd.   'bois malgache batard, bois savon'

Summerhayes, Gwynne, Stoddart (1),(2), Robertson(7)

Dup 17/33(K), Rob 2836,3024,3114(S)

Localities recorded: P,Pla,Coe,Dar,Alp

Indigenous slender widely branching shrub or tree with dark red stems or trunk, long-petioled ovate serrated edged leaves to 15 cm long and pale beneath, long axillary sprays of bead-like tight clusters of small greenish flowers and fleshy white globose fruit to 1 cm wide, the black seeds embedded in the flesh.

Procris Comm. ex Juss.

P. latifolia Blume

Summerhayes

Jef 507(S),577(K), Sch 11503(K), Lam 411(K), Fos 52022(S), Pro 4075(K), Rob 2480(S)

Localities recorded:; M,S

Indigenous succulent herb or subshrub with thick stems, elliptic serrated edged glossy green leaves, paler beneath, to 25 cm long, ♂ flowers small and white in axillary stalked clusters, ♀flowers in dense cushions on main stems.

ULMACEAE (153/2)

Trema Lour.

T. orientalis (L.)Blume (*T. guineensis*)   'bois malgache'

Summerhayes, Wilson (1)

Jef 520(S), Pro 3980(S), Rob 2543(S)

Localities recorded: M,S,P

Indigenous shrub or small tree to 8 m with softly hairy long pointed serrated edged triangular leaves to 12 cm long, two ranked and drooping on either side of the stem, axillary clusters of minute greenish white flowers and small globose fleshy fruit.

CANNABACEAE (153/4)

Cannabis L.

C. sativa L.   'chanvre indien, gandia, hemp, bhang, hashish, ganja, charas'

Bailey

No locality recorded

Erect annual herb to 5 m with hairy long-petioled palmately compound leaves with 3 to 11 leaflets, ♂ and ♀ flowers usually on separate plants, ♂ flowers yellow to 1 cm wide, ♀ flowers green, with brown shiny fruit to 5 mm long enclosed in bracts. (Fig. 161)

Fig. 161. Cannabis sativa

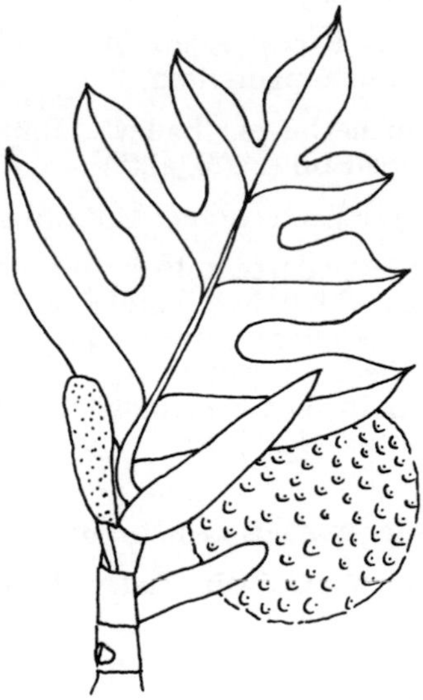

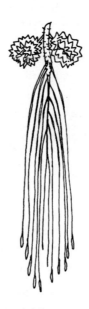

Fig. 162. Artocarpus altilis

Fig. 163. Ficus bojeri

Fig. 164. Casuarina equisetifolia

MORACEAE (153/5)

Artocarpus J.R. & G.Forst.

A. altilis (Park.)Fosberg (*A. communis, A. incisus*)   'breadfruit, breadnut, arbre à pain, fruit à pain, rima'

Summerhayes, Bailey, Lionnet(1),(2), Procter (2), Fosberg, Wilson,(2), Robertson (2),(3),(4),(5)

Localities recorded: M,P,F,N,C,Co,Coe,Poi

Cultivated tree to 15 m with very large deeply pinnately lobed leaves to 50 cm long with paired ♂ and ♀ receptacles, ♂ oblong to 15 cm long and ♀ globose and developing into the globose green fruit 20 cm wide and containing edible white flesh, yellow when mature, seedless in the breadfruit and with large ovoid brown seeds in the fertile form, the breadnut. The seeds can be eaten if cooked. (Fig. 162)

A. heterophyllus Lam. (*A. integrifolius*)   'jackfruit, jacquier, jak'

Sumner Gibson, Bailey, Lionnet (1), Procter (2), Wilson (2), Robertson (4),(5),(6)

Localities recorded: M(sr only),S,P,F,N,C,An

Cultivated and widely escaping tree to 15 m with ovate leaves to 15 cm long, the young plant with lobed leaves, separate ♂ and ♀ receptacles, ♂ terminal on short leafy shoots from the main trunk or large branches and to 10 cm long, ♀ below on stout twigs and developing into a very large ovoid yellow fruit covered with short tubercles, to 90 cm long and containing brown seeds each enclosed in a yellow edible fleshy covering and mixed with white sticky fibres, with a pungent sweet odour.

Chlorophora Gaudich.

C. excelsa (Welw.)Benth.   'mvuli'

Bailey

No locality recorded

Cultivated timber tree to 30 m with ovate pointed leaves to 15 cm long, the young leaves serrated edged, pendulous ♂ catkins to 15 cm long and ♀ spikes erect, stout, to 6 cm long developing into green oblong fruit composed of many achenes.

Ficus L.

F. avi-avi Blume (*F. consimilis*)   'multipliant petite feuille, lafouche petite feuille'

Summerhayes, Bailey, Renvoize, Fosberg (2)

Ves 6129(K), Jef 837(S), Fos 52143(S), Fra/Feare 458(S), Tod 28(S)

Localities recorded: M,S,P,F,Co,V,Ald,Ass,Cos,Ast

Indigenous tree with milky latex, ovate leaves to 12 cm long, the young leaves longer, stipules to 10 cm long or more, petioles to 3.5cm long and the lateral leaf veins at an angle of 80° from the midrib, and globose fruit to 8 mm wide on stalks to 4 mm long.

F. benghalensis L. 'banyan, multipliant,lafouche multipliant'

Summerhayes, Bailey, Gwynne, Lionnet (1), Renvoize, Stoddart (1),(3), Robertson (3),(6)

Jef 412(S), War 190(S), Rob 2668,3120(S)

Localities recorded: M,F,A,An,Bir,Den,Coe,?Des,Poi,Far

Huge tree with roots descending from the branches to form additional trunks, with milky sap, roundish leaves to 20 cm long and flattened globose sessile bright red fruit to 1.5 cm wide.

F. benjamina L. (*F. retusa* var. *nitida*, ?*F. terebrata*) 'ceylon willow, java fig, afouche or lafouche batard blanc'

Summerhayes

Localities recorded: M

Cultivated small tree with thin roots descending from the branches, pale green shining leaves with pointed tip to 10 cm long and solitary sessile fleshy purple fruits to 1 cm wide.

F. bojeri Baker

Summerhayes, Procter (1)

Ves 5518(K), Jef 432,1170(S), Pro 4074,3936,4369(S), Ber 14680(K), Rob 2632(S)

Localities recorded: M,S,D,Fe

Endemic small tree with milky sap, thin ovate pointed leaves with serrated edges and rough to touch to 20 cm long, with globose green fruits to 1 cm wide on stalks to 1 cm. (Fig. 163)

F. carica L. 'figuier, fig'

Bailey

Localities recorded: M(sr only)

Cultivated tree to 10 m with deeply 3–5-lobed leaves, scabrid, palmately veined, to 20 cm long, with green pear-shaped edible fruit.

F. elastica Roxb. 'rubber tree'

Robertson (7)

Localities recorded: M(sr only),Dar

Large tree with radiating buttress roots and roots descending from the branches, but usually grown as an ornamental pot plant with dark green glossy ovate pointed leaves to 25 cm long.

F. nautarum Baker   'la fouche grand feuille, afouche or lafouche rouge'

Summerhayes, Renvoize, Fosberg (2), Stoddart (1),(3), Wilson (1),(2), Robertson (4),(6)

Jef 1186(S), Fos 52142(S), Bru/War 69(S) (*F.lutea* Vahl at Kew), Rob 2858(S)

Localities recorded: M,S,P,F,C,A,An,Bir,Rem,Alp,Ald,Ass,Cos,Ast

Indigenous shrub or spreading tree to 10 m with elliptic oblong or obovate leaves to 30 cm long, milky sap, sessile green, tinged with pink, fruit to 1.8 cm wide. The tree is deciduous.

F. pumila L.   'creeping fig'

Localities recorded: M(sr only)

Cultivated creeper, usually on walls, with small sessile young leaves, heart shaped to 2 cm long, and large elliptic leaves to 10 cm long on the fruiting stems, with green fruit to 8 cm long.

F. reflexa Thunb. (*F. rubra* var. *sechellensis*, *F. sechellarum*)   'afouche or lafouche rouge, afouche, lafouche'

Summerhayes, Bailey, Renvoize, Fosberg (2), Evans

Ves 5765(K), Fos 52084,52134(S), Pro 4252(K), Fra/Feare 457(S), Bru/War 51(S), Rob 2657(S)

Localities recorded: M,S,P,F,A,Co,V,Ald,Ass,Cos,Ast

Shrub or small tree with milky juice, stalked ovate leaves to 15cm or more long with the lateral veins at 50° to the midrib, sessile globose fruit to 7mm wide, yellow when ripe.

Maillardia Frap. & Duch.

M. pendula Fosberg

Renvoize, Fosberg (2)

Localities recorded: Ald

Endemic slender tree to 10 m with pendulous branches and elliptic leaves to 9 cm long with long points, globose clusters of ♂ flowers, yellowish, and reddish brown ovoid fruit to 1.5 cm long, containing one grooved seed.

Morus L.

M. indica L.   'murier, mulberry, indian mulberry'

Summerhayes, Bailey (probably not *M. nigra*)

Localities recorded: M

Cultivated shrub or small tree with heart-shaped or lobed serrated edged leaves to 20cm long, and cylindrical juicy edible black fruits composed of many drupes. (The nomenclature of *Morus* spp. appears to be confused)

Trilepisium Thouars

T. madagascariense DC. (*Bosquiea gymnandra*)

Summerhayes, Procter (1)

Localities recorded: M,S

Tree with milky latex, ovate leaves with elongated points to 15 cm long, stipules to 1 cm long, and axillary many-stamened white flowers, each to 1 cm wide. (Procter says not endemic)

CASUARINACEAE (158)

Casuarina L.

C. equisetifolia L. (*C.littorea*)   'cedre, filao, pin, whistling pine'

Summerhayes, Sumner Gibson, Bailey, Gwynne, Lionnet (1), Sauer, Renvoize, Procter (2), Fosberg (2), Stoddart (1),(2),(3), Wilson (4), Piggott, Robertson (4),(5),(6)

Jef 1195(S), Fos 52168(S), Bru/War 84(S), Rob 2384,2848,3007, 3080(S)

Localities recorded: M,S,P,F,N,C,A,An,Co,Bir,Den,Pla,Coe,Rem,Des,Dar, Jos,Poi,Mar,Alp,Pro,Far,Ald,Ass, Cos, Ast

Tree to 20 m with slender green cylindrical jointed branchlets acting as leaves, true leaves scale like at nodes, ♂ flowers in narrow terminal catkins, ♀ flowers in axillary blunt spikes which develop into globose cone-like false fruit containing the flattened winged fruits, each to 7 mm long. (Fig. 164)

M O N O C O T Y L E D O N S

HYDROCHARITACEAE (167)

Enhalus Rich.

E. acoroides (L.f.)Rich. ex Steud.

Summerhayes

Jef 1168(S)

Localities recorded: M,P

Marine 'grass' with thick white roots, rhizomes covered with the stiff wiry remains of the leaf sheaths and long strap-like leaves to 90 cm long.

Halophila Thouars

H. minor (Zoll. ex Miq.)den Hartog (*H. ovalis*)

Summerhayes, Fosberg (2)

Jef 496(S)

Localities recorded: M,L,Ald

Marine 'grass' with thin fragile rhizomes and ovate long-stalked leaves, the blade to 2 cm long.

Thalassia Koenig

T. hemprichii (Ehrenb.)Aschers.   'turtle grass'

Summerhayes, Bailey, Gwynne, Renvoize, Fosberg (2), Stoddart (1),(2)

Localities recorded: M,L,Mo,Dar,Jos,Far,Ald,Cos

Marine 'grass'with thin roots, rhizomes to 3 mm wide and narrow curved strap-like leaves to 40 cm long, but often shorter, with papery fibrous sheaths. (Fig. 165)

ORCHIDACEAE (169)

(Note — the status of many of the endemic orchids is now in some doubt)

Acampe Lindl.

A. praemorsa (Roxb.)Blatter & MacCann (*A. pachyglossa* ssp. *renschiana*, *A.rigida*)   'fleur de putantil'

Fosberg (2)

Arc 168(K)

Localities recorded: P,Ald

Epiphytic orchid with long aerial roots, leathery leaves in two ranks, folded, to 22 cm long, and axillary racemes to 7 cm long of fragrant flowers to 1.2 cm long, yellow blotched with reddish purple, and long narrow capsules to 5 cm long.

Agrostophyllum Blume

A. occidentale Schltr.

Summerhayes, Bailey, Procter (1),(3)

Arc 182(K), Jef 732(K), Pro 4020(K), Rob 2481(S)

Localities recorded: M,S

Endemic epiphytic orchid with stems to 30 cm, leaves in two ranks, base folded and tight to stem, blades flat to 15 cm long, small pale yellow flowers to 5 mm long in dense terminal rounded sessile clusters 5 cm wide, and capsules to 6 mm long.

Angraecum Bory

A. eburneum Bory (including ssp. *superbum* (Th.)H.Perr.) (*A. brongniartianum*) 'paille-en-queue, tropic bird orchid'

Summerhayes, Bailey, Piggott, Lionnet (1),(2), Procter (3), Fosberg (2)

Jef 737(S),474,694,1192(K)

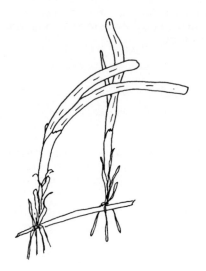

Fig. 165. Thalassia hemprichii

Fig. 166. Bulbophyllum intertextum

Fig. 167. Malaxis seychellarum

Fig. 168. Polystachya concreta

Localities recorded: M,S,P,Far,Ald

Indigenous robust epiphytic orchid with oblong leathery thick leaves to 60 cm long in two ranks, the base folded, with erect axillary racemes to 75 cm long bearing widely spaced large white or yellowish green flowers to 5 cm long with a spur to 15 cm long, and capsules to 5 cm long.

A. zeylanicum Lindl. (*A. maheense*)

Summerhayes, Bailey, Lionnet (1), Procter (1),(3)

Arc 174(K)

Localities recorded: M

Epiphytic orchid (at one time thought to be endemic) with two ranked oblong pointed leaves to 20 cm long by 2 cm wide, and slender arching racemes to 10 cm long with widely spaced small star-shaped (?)yellow flowers to 8 mm wide with short fat spur as long as the sepals.

Arachnis Blume

A. × maingayi (Hook.f.)Schltr.   'scorpion orchid'

Bailey

No locality recorded

Cultivated epiphytic orchid with long climbing stems branching where bent, narrow oblong leaves to 15 cm long, bilobed at the tip, and racemes of large striking flowers, the three sepals and two petals long and narrow, white with pink blotches and margins, and the lip pink.

Bulbophyllum Thouars

B. intertextum Lindl. (*B. seychellarum*)

Summerhayes, Bailey, Procter (1),(3)

Arc 176(K), Jef 546,794(S), 765(K)

Localities recorded: M,S

Indigenous (thought previously to be endemic) small epiphytic orchid with globose green pseudobulbs to 8 mm wide, each with one erect oblong leaf to 3 cm long, and basal slender erect racemes to 20 cm long, with greenish white and purple star-shaped flowers to 5 mm wide and subglobose capsules to 5 mm wide. (Fig. 166)

Calanthe R.Br.

C. triplicata (Willem.)Ames (*C. ?sylvatica*)

Summerhayes, Bailey, Procter (3)

Arc 180(K), Jef 575(S),1214(K), Pro 4566(K)

Localities recorded: M

Indigenous ground orchid to 75 cm tall with pseudobulbs, tufts of large green pleated leaves to 10 cm wide and erect terminal racemes of large white flowers, each to 2 cm long with a spur a little longer than the flowers, and drooping ovoid capsules to 4 cm long.

Cattleya Lindl.

C. × Mantinii (*bowringiana* × *dowiana*)

Bailey

No locality recorded

Cultivated showy epiphytic orchid with slender pseudobulbs and stiff fleshy leaves with clusters of large flowers, each to 15 cm long, with mauve sepals and purplish petals, the trumpet-shaped lip deeper purple with yellow in the throat.

Cirrhopetalum Hook. & Arn.

C. umbellatum (Forst.)Hook.& Arn. (*C. thouarsii, Bulbophyllum longiflorum*)

Thomasset n/n,100(K), Horne 597,598(K), Gardiner n/n(K) (Summerhayes was unable to place these), Arc 172(K), Jef 548(K)

Localities recorded: M

Indigenous epiphytic orchid with smooth pseudobulbs from a creeping rhizome each bearing a single leathery leaf to 15 cm long, a slender inflorescence to 20 cm long bearing flowers with two long tawny orange sepals to 2.5 cm long and small yellow petals.

Cynorkis Thouars

C. fastigiata Thouars

Summerhayes, Bailey, Procter (3), Wilson (2)

Arc 179(K), Jef 361(S), Rob 2279(S)

Localities recorded: M,S,P,C

Indigenous tuberous rooted ground orchid with two slightly fleshy ovate oblong leaves flat to the ground and an erect raceme to 30 cm tall with one to few flowers at the tip, each flower pale yellow with mauve marks, the lip obvious and 4 lobed, to 1cm long, and a spur to 2.5 cm long, and the capsule narrow to 3 cm long.

Dendrobium Sw.

D. crumenatum Sw.   'pigeon orchid'

Bailey, Procter (3)

Localities recorded: M(sr only)

Cultivated and widely naturalised epiphytic orchid with furrowed yellow pseudobulbs and erect stems with leathery ovate leaves to 7 cm long and delicate white flowers to 2 cm long, which from above look like a bird in flight, lasting one day. The plants all flower together 8 to 11 days after a drop in temperature.

D. phalaenopsis Fitzger.

Bailey

No locality recorded

Cultivated epiphytic orchid with pseudobulbs to 60 cm long with stiff pointed leaves to 25 cm long and long stalked terminal racemes with 6 to 15 large flat flowers 6 to 8 cm wide in two rows with wide ovate perianth segments and a narrow oblong lip.

D. pulchellum Roxb. (*D. dalhousieanum*)

Bailey

No locality recorded

Cultivated epiphytic orchid to 1.5 m with hard woody pseudobulbs, lanceolate leaves to 25 cm long and racemes from leafless pseudobulbs with 3 to 10 large flat flowers, each to 2 cm wide, with yellow and pink petals and a downy cup-shaped lip.

Disperis Sw.

D. tripetaloides (Thouars)Lindl. (*D.cordata*)

Summerhayes, Bailey, Procter (3)

Arc 181(K), Jef 407(S),812(K), Pro 4010(K)

Localities recorded: M,S,P

Indigenous tuberous-rooted ground orchid with erect racemes to 20 cm tall bearing alternate heart-shaped leaves, the lowest usually flat on the ground, rich green mottled brown, and small white flowers with mauve markings in the hooded sepal, to 1 cm wide, with narrow capsules to 2 cm long.

Epidendrum L.

E. radicans Lindl.

Bailey

No locality recorded

Cultivated epiphytic orchid with long thin stem-like pseudobulbs to 1 m or more with many long wiry aerial roots, fleshy oblong leaves to 10 cm long and terminal pyramidal racemes of flowers with narrowly ovate perianth segments and a fringed lip to 2 cm wide, of various colours, white, orange, yellow, pink or mauve, and drooping ovoid capsules to 4 cm long

Grammatophyllum Blume

G. speciosum Blume   'giant orchid, queen of orchids'

Bailey, Lionnet (2)

Localities recorded: M(BG)

Cultivated epiphytic orchid with very large pseudobulbs to 3 m tall with stout oblong two ranked leaves to 75 cm long and inflorescences to 2 m tall with flowers to 15 cm wide, yellow blotched with reddish purple.

Graphorkis Thouars

G. scripta (Thouars)Kuntze (*Eulophia sp.*)

Summerhayes (Gardiner 101)

Localities recorded: S

Indigenous epiphytic orchid with yellow ribbed pseudobulbs with several nodes and fibrous remains of leaves, tufted strap-shaped leaves to 50 cm long and pyramidal panicles, from the base of the pseudobulb, of wide petalled flowers, green with brown spots, and a short spur, to 1.5 cm wide, with ovoid capsules to 2.5 cm long.

Hederorkis Thouars

H. seychellensis Bosser (*Bulbophyllum scandens*)

Summerhayes, Fosberg (2), Procter (1),(3)

Arc 173(K)

Localities recorded: M,S,?Ald

Endemic robust epiphytic orchid with squarish stems to 2 m long with ovate oblong leathery 2 ranked leaves to 17 cm long and up to 7 cm wide, and curving racemes to 10 cm long with cream or purple flowers, the segments to 1.2 cm long, and ovoid capsules to 2 cm long.

Liparis L.C.Rich.

L. flavescens (Kuntze)Lindl.

Summerhayes, Bailey, Procter (3)

No locality recorded

Indigenous epiphytic (? or ground) orchid with narrow pseudobulbs and basal wide ovate pointed leaves, the blade to 12 cm long and 4 cm wide, with erect racemes of yellowish flowers, each to 1 cm long,and ovoid erect capsules to 1 cm long.

Malaxis Sol. ex Sw.

M. seychellarum (Kraenzl.)Summerhayes (*Microstylis seychellarum*)

Summerhayes, Bailey, Procter (1),(3)

Arc 170(K), Jef 452,456,552(S),401(K), Sch 11674(K), Pro 4021,4512(K), Rob 2634(S)

Localities recorded: M,S

Endemic epiphytic orchid (often on the ground) with pseudobulbs and wide ovate leaves to 20 cm long and 7 cm wide with erect racemes of small yellow or purple or green flowers, each to 4 mm long and often densely arranged, with obovoid erect capsules to 1 cm long. (Fig. 167)

Oeceoclades Lindl.

O. seychellarum (Rolfe ex Summerhayes)Garay & Taylor (*Eulophia seychellarum, Eulophidium seychellarum*)

Summerhayes, Bailey, Procter (1),(3)

Localities recorded: M(? no longer)

Endemic epiphytic orchid with creeping stem, pointed ovate leaves to 12 cm long in pairs, slender racemes to 30 cm long bearing delicate white flowers on stalks, each flower to 8 mm long with purple marks and a short spur, and narrow capsules to 2 cm long.

Oeoniella Schltr.

O. aphrodite Schltr.

Summerhayes, Procter (3)

Arc 169(K)

Localities recorded: M,P,S

Indigenous epiphytic orchid with branching stems, many aerial roots and short leathery oblong two ranked leaves to 25 cm long and bilobed at the tip, with slender axillary racemes of spurred white flowers each to 1.5 cm wide, the lip bilobed with a pointed 'tongue', and oblong capsules to 2 cm long.

Oncidium Sw.

O. sphacelatum Lindl.

Bailey

No locality recorded

Cultivated epiphytic orchid with ovate, slightly compressed pseudobulbs to 25 cm long, bearing 2 to 3 leaves to 70 cm long and panicles to 2 m long with showy yellow, marked with brown, flowers to 2.5 cm wide.

Orchis L.

O. purpurea Huds.   'lady orchid'

Bailey

No locality recorded

Cultivated ground orchid to 1m with tuberous roots, wide lanceolate basal leaves to 25 cm long and erect racemes of densely arranged dark purplish flowers, each to 1 cm wide with a pale lip. There must be considerable doubt that this orchid was ever grown in Seychelles.

Phaius Lour.

P. tetragonus (Thouars)Reichb.f.

Summerhayes, Bailey, Lionnet (1), Procter (3)

Localities recorded: M

Indigenous ground orchid with large ovate folded leaves to 30 cm long and 10 cm wide and tall few-flowered racemes of white flowers with narrow perianth segments and large lip but no spur.

Platylepis A.Rich.

P. goodyeroides A.Rich. (*P. occulta*)

Summerhayes, Bailey, Procter (3)

Arc 177(K), Jef 458(K), Pro 4486(K)

Localities recorded: M,S

Indigenous trailing ground orchid with long stems and widely ovate petioled leaves to 10 cm long and 4 cm wide, with erect racemes of many densely arranged small pink flowers with white bracts and white capsules to 8 mm long.

P. sechellarum S.Moore

Summerhayes, Bailey, Procter (1),(3)

Arc 177A(K), Pro 4570(K)

Localities recorded: M,S

Endemic trailing ground orchid with narrower leaves than *P. occulta* and slender racemes with widely spaced small light purple flowers, each to 4 mm long.

Polystachya Hook.

P. concreta (Jacq.)Garay & Sweet (*P.bicolor, P.mauritiana, P.zeylanica*)

Summerhayes, Bailey, Procter (3)

Arc 183(K), Jef 1223(S),547,783,1213,513,555,700(K)

Localities recorded: M,S.

Indigenous epiphytic orchid with basal obovoid oblong leaves to 20 cm long, folded at the base, and many roots, pseudobulbs not swollen, and terminal racemes with long stem clasping bracts and short side branches, all on one side, bearing the 5 mm wide hood-shaped pink or yellow flowers with a white lip, followed by narrow capsules to 1.5 cm long. (Fig. 168)

P. fusiformis (Thouars)Lindl.

Summerhayes, Procter (3)

Localities recorded: M

Indigenous epiphytic orchid with long, branched stems, usually hanging down, narrower leaves than *P.concreta*, and inflorescences with branched side branches and flowers to 3 mm wide.

Spathoglottis Blume

S. plicata Blume   'palm orchid'

Bailey, Procter (3)

Localities recorded: M(sr only)

Cultivated ground orchid which has often escaped, with clustered pseudobulbs to 4 cm tall, basal lanceolate leaves to 70 cm tall and 10 cm wide and axillary erect racemes to 80 cm with up to 25 flowers with pinkish bracts, the flowers pink, violet or white with a 3-lobed lip, to 4 cm wide. (Fig. 169)

Vanda R.Br.

V. teres (Roxb.)Lindl.

Bailey, Lionnet (2)

Localities recorded: M(BG)

Cultivated epiphytic orchid with branching tangled green rounded stems and stem-like green leaves, alternate and angled to the stems, to 20 cm long, with axillary racemes of up to 6 large pinkish purple flowers, to 10 cm wide, on twisted stems, with a 3-lobed lip and short spur.

Vanilla Sw.

V. phalaenopsis Reichb.f.   'vanille sauvage'

Summerhayes, Bailey, Lionnet (1),(2), Procter (1),(2),(3),

Jef 731(S),375(K), Sch 11692(K), Fos 51988(K), Rob 2546(S)

Localities recorded: M,S,P,Fe

Endemic epiphytic orchid with trailing rounded green stems, leafless except at the apex, short clasping roots and stout racemes of large white flowers to 8 cm long with a peach-coloured trumpet-shaped lip, one flower opening at a time and fading by midday, and fat green curved cylindrical pods to 12 cm long.

V. planifolia Andrews (*V. fragrans, V. mexicana*)   'vanilla'

Bailey, Gwynne, Lionnet (1), Procter (2),(3), Stoddart (2), Wilson (2), Robertson (2),(4)

Rob 3199(S)

Localities recorded: M,P,F,C,Coe,Dar

Cultivated epiphytic orchid which has often escaped, with rounded green stems, often many metres long, and fleshy oblong leaves to 10 cm long, alternately arranged, with axillary racemes to 10 cm long of up to 20 large greenish white flowers opening singly. The flowers must be hand pollinated to produce the pods which yield vanillin.

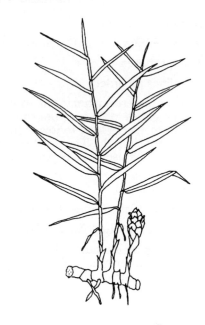

Fig. 169. Spathoglottis plicata

Fig. 170. Zingiber officinale

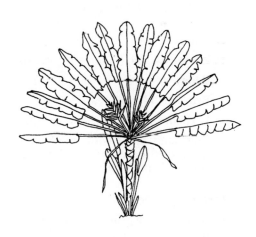

Fig. 171. Maranta arundinacea

Fig. 172. Ravenala madagascariensis

ZINGIBERACEAE (170/1)

Aframomum K.Schum.

A. angustifolium (Sonn.)K.Schum.   'wild or madagascar cardamon'

Summerhayes

Localities recorded: M

Indigenous erect plant to 3 m with two ranked leaves to 30 cm long and erect flower spikes to 30 cm from the underground stems, the flowers pink and yellow to 4.5 cm long followed by red globose fruits containing tiny brown seeds in acid pulp. The crushed leaves have a spicy ginger odour and the seeds can be used as a pepper substitute.

Alpinia Roxb.

A. zerumbet (Pers.)Benth. & Smith   'shell ginger'

Localities recorded: M(sr only)

Cultivated rhizomatous plant to 4 m with two ranked ovate lanceolate leaves and arching terminal spikes of fragrant flowers, the bracts waxy white and the perianth whitish with a red and yellow lip.

Amomum L.

A. magnificum Benth. & Hook. (*Phaeomaria magnifica*)   'langouze, torch ginger'

Summerhayes, Bailey, Lionnet (2)

Localities recorded: M(BG)

Cultivated rhizomatous plant to 2.5 m tall with two ranked lanceolate leaves to 30 cm long and long-stalked ornamental conical pink heads to 15 cm wide, each cone of flowers with a basal circle of stiff pink bracts.

Curcuma L.

C. domestica Val. (*C. longa*)   'safran, turmeric, curry plant'

Bailey

Bru/War 110(S)

Localities recorded: M(sr only),A

Cultivated plant with small root tubers, rhizomes with yellow flesh and lanceolate leaves to 30 cm long, and yellow and white flowers to 5 cm long in coloured cone-like terminal shoots. The yellow rhizome yields the spice.

Elettaria Maton.

E. cardamomum Maton.   'cardamon'

Bailey

Jef 448(S), 612(K), Rob 2536(S)

Localities recorded: M

Cultivated and widely escaping plant with clumps of erect leafy stems to 5 m tall with two ranked lanceolate leaves to 90 cm long and a long drooping spike to 1.2 m long from the rootstock with greenish and mauve streaked flowers to 1 cm long and trilocular capsules containing up to 20 aromatic seeds. The dried capsules are the spice.

Hedychium Koenig

H. coronarium Koenig   'mon ame, ginger lily'

Summerhayes

Localities recorded: M

Cultivated rhizomatous plant with two ranked lanceolate leaves to 50 cm long and terminal clusters of large fragrant white flowers each to 10cm long, with a yellow spot on the lip.

H. flavescens Carry. ex Roscoe

Localities recorded: M(sr only)

Cultivated and escaping rhizomatous plant similar to *H. coronarium* but with fragrant pale yellow flowers and taller stems.

Zingiber Boehm.

Z. officinale Roscoe   'gingembre, ginger'

Bailey

Bru/War 111(S)

Localities recorded: M(sr only),A

Cultivated rhizomatous herb, with aromatic rhizomes with pale yellow flesh, annual two ranked leafy shoots to 50 cm, the leaves linear lanceolate to 25 cm long, and a rarely seen cone-like spike from the rootstock to 7 cm tall, the yellowish bracts enclosing the yellow and purple flowers. The dried rhizome yields the spice and is preserved in syrup as a sweetmeat. (Fig. 170)

Z. zerumbet (L.)Smith   'wild ginger'

Localities recorded: M(sr only)

Herb from rhizomatous rootstock with two ranked leafy stems to 2 m, the leaves broadly lanceolate to 30 cm long, and a cone-like spike to 20 cm from the rootstock with red bracts holding water and the 5 cm long cream-coloured flowers.

COSTACEAE (170/2)

Costus L.

C. speciosus (Koenig)Smith   'spiral ginger'

Localities recorded: M(sr only)

Cultivated perennial with rhizomatous rootstock, slender spiral stems to 3m with ovate lanceolate leaves and dense terminal clusters of red bracts and large white flowers, each to 7cm wide.

MARANTACEAE (170/3)

Calathea G.F.W.Mey.

C. allouia (Aubl.)Lindl.   'topi nambour'

Bailey

No locality recorded

Cultivated perennial herb to 1m with clumps of *Canna*-like elliptic leaves to 60 cm long, greyish beneath, with white flowers, each to 2.5 cm long, and small edible ovoid tubers at the ends of fibrous roots.

Maranta L.

M. arundinacea L.   'arouroute, arrowroot'

Bailey,Stoddart (3)

Bru/War 63(S), Rob 2623(S), Tod 59(S)

Localities recorded: M(sr only),F,N,A,Bir

Cultivated erect glabrous perennial to 1.5 m tall, with edible white fleshy cylindrical rhizomes to 5 cm wide and 40 cm long with overlapping brown scale leaves, ovate leaves to 15 cm long and terminal spikes with paired white flowers, each to 2 cm long. (Fig. 171)

M. spp. (incl. *M. bicolor* Ker. and *M. leuconeura* Morr.)   'maranta'

Bailey

Localities recorded: M(sr only)

Ornamental plants, similar to *M.arundinacea* but with variegated leaves.

MUSACEAE (170/4)

Heliconia L.

H. bihai L.   'lobster claw'

Robertson (7)

Localities recorded: M(sr only),Dar

Ornamental perennial with long-petioled oblong leaves to 3 m tall and terminal erect flattened spike of several bright crimson bracts with green edges, each bract to 12 cm long, containing the greenish flowers.

H. brasiliensis Hook.   'mordant crabe, lobster claw'

Bailey, Lionnet (2)

Localities recorded: M(BG)

Ornamental plant similar to *H. bihai* but with scarlet bracts and pink to purple flowers. (These two species are often confused)

H. psittacorum L.f. 'parrot flower'

Localities recorded: M(sr only)

Ornamental perennial with long stemmed lanceolate leaves to 1.5 m tall and long stalked spikes of a few narrow orange bracts containing erect dark tipped flowers.

H. rostrata Ruiz & Pav. 'hanging lobster claw'

Localities recorded: M(sr only)

Ornamental perennial similar to *H. bihai* but with the spike of bracts hanging down.

Musa L.

M. spp. (incl. *M. paradisiaca* L. and *M. sapientum* L.) 'bananier, banana, plantain'

Bailey, Lionnet (1), Fosberg (1),(2), Stoddart (1),(2),(3), Baumer, Wilson (1),(2),(4), Robertson (2),(3),(4),(5)

Localities recorded: M(sr only),P,F,N,C,Co,Bir,Den,Coe,Des,Dar,Poi, Mar,Far,Ass,A(sr only)

Widely cultivated tree-like perennial herb to 5 m with basal corm, false stem of fleshy petioles and oblong leaves, and a terminal downwards curving spike with several wide purplish bracts containing a row of flowers which develop into oblong curved edible fruit ripening green, yellow or red. The plantain usually curves upwards and the sweet banana curves downwards. In general sweet bananas are *M. sapientum* and plantains are *M. paradisiaca,* but the nomenclature is confused. There are at least 25 named varieties in Seychelles and some are used medicinally as well as for food.

Ravenala Adans.

R. madagascariensis J.F.Gmel. 'banane voyageur, palmier de voyageur, traveller's palm'

Bailey, Lionnet (2), Procter (2).

Localities recorded: M(BG),P

Ornamental tree with fan-shaped arrangement of banana-like leaves and stout spike of whitish bracts and flowers. (Fig. 172)

Strelitzia (Banks)Ait.

S. reginae Ait. 'bird of paradise, crown bird flower'

Localities recorded: M(sr only)

Ornamental plant with clumps of long-petioled oblong glaucous leaves to 1.5 m tall with striking long stalked spikes with pointed greyish green bracts containing erect flowers with pointed orange and bright blue perianth segments like a bird's crest and beak.

CANNACEAE (170/6)

Canna L.

C. indica L. (incl. *C. edulis*) (*C. bidentata*)   'balisier, canna, safron marron, tous les mois'

Summerhayes, Bailey

Bru/War 73(S)

Localities recorded: M,A,Bir

Indigenous erect herb to 1.5 m with stout rhizomes, large ovate leaves to 30 cm long, terminal spikes of red flowers to 4 cm long and globose softly spiny capsules to 3 cm wide containing hard black globose seeds to 6 mm wide.

C. hybrids

Robertson (7)

Localities recorded: M(sr only),Dar

The modern ornamental garden *Canna* is derived from crosses between various *Canna* spp. including *C. glauca* and *C. discolor*. Similar to *C. indica* but with showy brightly coloured flowers to 15 cm long.

BROMELIACEAE (171)

Ananas Mill.

A. comosus (L.)Merr. (*A. sativus*)   'ananas, ananas marron, pineapple'

Summerhayes, Bailey, Lionnet (1), Fosberg (1), Procter (2), Wilson (2), Robertson (4),(5),(6)

Localities recorded: M(sr only),P,F,N,C,An,Co,A(sr only)

Cultivated perennial herb to 1 m with a rosette of stiff lanceolate prickly-margined yellowish green leaves with a compact terminal swollen spike of purplish blue flowers which develops into the cylindrical yellow fleshy edible fruit topped by a small crown of leaves. A wild form occurs which has brilliant red flower bracts and may be *A. bracteatus*.

IRIDACEAE (173/1)

Gladiolus L.

G. spp.   'glaieul'

Bailey

Localities recorded: M(sr only)

Ornamental plant with corms, erect sword like leaves to 1.5 m tall and large striking flowers on a one sided spike, the flower to 7 cm long, funnel shaped with the upper 3 segments larger than the lower 3.

Neomarica Sprague

N. northiana (Schneev.) Sprague   'walking iris'

Localities recorded: M(sr only)

Ornamental perennial to 1 m with sword-shaped leaves to 75 cm long and 5 cm wide, with leaf-like flower stalks bearing a few fragrant white flowers with brown and violet inner segments, to 6 cm wide, the flower stalks rooting where the flowers were.

AMARYLLIDACEAE (174/1)

Crinum L.

C. amabile Ker-Gawl. (*C.augustum*)   'lis du pays'

Summerhayes, Bailey, Gwynne, Lionnet (1), Stoddart (2), Robertson (4),(6)

Localities recorded: M,?F,C,?An,Coe,Des,Dar,Jos,Far

Bulbous-rooted plant with strap-shaped leaves to 1 m long and a flattened flower stalk to 1 m long bearing a cluster of many curved trumpet-shaped flowers, each to 10 cm long, red in bud and then red outside, white inside.

C. asiaticum L.   'lis blanc'

Summerhayes, Bailey, Lionnet (1), Fosberg (1)

Localities recorded: M,Co

Bulbous-rooted plant with wide lanceolate leaves to 1.5 m tall and stout stalk bearing a cluster of many large white flowers with narrow perianth segments, each to 10 cm long with purple filaments and orange anthers.

C. jagus (Thomps.)Dandy (*C. giganteum*)

Localities recorded: M(sr only)

Bulbous-rooted plant with strap-shaped leaves to 1 m with a flattened stem to 1 m bearing a few-flowered cluster of white bell-shaped flowers, each to 18 cm long and very fragrant.

C. macowanii Baker

Rob 3131(S)

Localities recorded: M(sr only),Coe

Bulbous-rooted plant to 2 m tall with stout stem bearing large lanceolate leaves and long-stalked cluster of many red and white striped trumpet-shaped flowers to 15 cm long.

C. moorei Hook.f.   'lis rose'

Bailey, Lionnet (1)

No locality recorded

Bulbous-rooted plant with lanceolate leaves with wavy margins, long-stalked cluster of 6 to 12 large pink flowers, tinged with white at the base.

C. zeylanicum (L.)L.   'lis st joseph'

Summerhayes, Bailey

Localities recorded: M

Bulbous-rooted plant with lanceolate leaves to 1 m long and long-stalked clusters of up to 20 pendulous white flowers with a purplish pink stripe down each petal.

Eurycles Salisb.

E. sylvestris Salisb.

Localities recorded: M(sr only)

Bulbous-rooted plant with wide ovate or rounded leaves to 15 cm wide and 20 cm long with long-stalked heads of white flowers, each flower with a cup-shaped structure at the base of the perianth segments.

Hippeastrum Herb.

H. puniceum (Lam.)Urb. (*H.equestre*)   'lis rouge, amaryllis, fire lily'

Bailey, Baumer

No locality recorded

Bulbous-rooted plant flowering before the strap-shaped leaves appear, with long-stalked cluster of up to 4 flowers, each flower held at right angles to the stem, to 12cm wide, red with a greenish centre. Used medicinally in Seychelles.

Hymenocallis Salisb.

H. littoralis (Jacq.)Salisb. (*H. caribaea*)   'lis belle de nuit, spider lily, filmy lily'

Bailey, Sauer, Fosberg (1),Stoddard (2), Robertson (7)

Localities recorded: M(sr only),Co,Dar,Jos

Bulbous-rooted plant with strap-shaped leaves to 75 cm long and stout-stalked heads of white flowers, each stalked and to 15 cm wide with a funnel-shaped central part and 6 long narrow perianth segments and long stamens.

Lycoris Herb.

L. radiata Herb. (*Nerine japonica*)   'lis rouge petit, red spider lily'

Bailey

No locality recorded

Bulbous-rooted herb to 45 cm tall with linear leaves which fade before the
flowers appear, and long stemmed heads of bright pink flowers to 10 cm long
with narrow strap-shaped petals which curl backwards from the long erect
stamens.

Scadoxus Raf.

S. multiflorus (Martyn.)Raf. (*Haemanthus multiflorus*) 'lis corail, blood lily,
painters brush lily'

Bailey

Rob 3053(S)

Localities recorded: Pla

Bulbous-rooted plant with ovate lanceolate leaves with wavy margins to 25 cm
long, and a stout red mottled flower-stalk bearing a rounded head of many
slender scarlet flowers, each to 4 cm long with long scarlet stamens. (Fig. 173)

Zephyranthes Herb.

Z. candida Herb.

Localities recorded: M(sr only)

Ornamental low growing bulbous-rooted plant with narrow or linear leaves
and solitary white flowers to 5 cm long on stalks as long as the leaves.

Z. citrina Baker   'crocus'

No locality recorded

Similar to *Z. candida* but with linear leaves and yellow flowers.

Z. rosea Lindl.   'lis rose petit'

Bailey, Robertson (3)

Localities recorded: M(sr only),Poi

Similar to *Z. candida* but with strap-shaped leaves and pink flowers. (Fig. 174)

HYPOXIDACEAE (174/2)

Curculigo Gaertn.

C. rhizophylla (Baker)Dur. & Schinz

Summerhayes, Procter (1)

Pro 4272(S), Rob 2773(S)

Localities recorded: M,S,P,D,C

Endemic plant with lanceolate, all basal leaves to 1.5 m long, often rooting at
the leaf tip, with shortly stalked flower spike at the leaf bases of a few dark
purple foetid flowers, each to 2.5 cm long.

C. sechellensis Boj.   'coco marron'

Summerhayes, Bailey, Lionnet (1), Procter (1)

Jef 484(S), Rob 2403(S)

Localities recorded: M,S,P,F

Endemic robust plant with all leaves basal, broadly lanceolate with split tips and spiny petioles and to 2 m long, and clusters of pale yellow flowers, each to 3 cm long, hidden among the leaf bases.

C. sp.

Summerhayes, Procter (1)

Ves 5479(K), Jef 701(S), Pro 4485,4516(K)

Localities recorded: M

Endemic plant, not yet described, similar to *C. rhizophylla* so that determinations quoted may be incorrect.

AGAVACEAE (174/4)

Agave L.

A. sisalana (Perr. ex Engelm.) Drumm. & Prain (*A.americana, A.rigida* var. *sisalana*) 'agave, sisal'

Summerhayes, Bailey, Lionnet (1), Fosberg (2), Stoddart (1),(3), Wilson (1),(2), Robertson (2),(4),(7)

Rob 2876(S)

Localities recorded: M,P,F,C,Bir,Den,Coe,Rem,Des,Dar,Alp,Far,Ald,Cos,Ast

Cultivated and widely escaping perennial plant with basal rosette of stiff spine tipped lanceolate glaucous leaves to 1.5 m long, with, after several years, a tall terminal spike (pole) to 7 m with clusters of erect creamy green flowers followed by bulbils which drop and take root. The leaf yields the fibre.

Furcraea Vent.

F. foetida (L.)Haw. (*F. gigantea*) 'agave gros, mauritius hemp'

Bailey, Gwynne, Lionnet (1), Fosberg (1), Wilson(4), Robertson (2),(3),(5)

Jef 1212(S), Rob 2537,2656,2842(S)

Localities recorded: M,F,N,Co,Coe,Poi,Mar,Alp

Cultivated and widely escaping plant with large basal rosette of lanceolate leaves to 3 m long and a tall terminal spike to 5 m with pendant creamy green flowers, each to 4 cm long. Grown for fibre from the leaf.

Polianthes L.

P. tuberosa L.   'tubereuse'

Bailey

Fig. 173. Scadoxus multiflorus

Fig. 174. Zephyranthes rosea

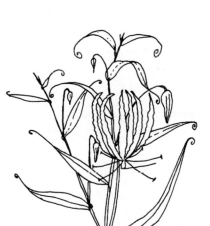

Fig. 175. Gloriosa superba

Fig. 176. Sansevieria trifasciata

No locality recorded

Cultivated tuberous-rooted plant with grass like leaves and waxy white fragrant flowers in a pyramidal stalked spike.

Yucca L.

Y. aloifolia L.   'yucca'

Bailey, Robertson (2)

Jef 590(S)

Localities recorded: M,Coe

Ornamental plant with almost glabrous rosette of stiff lanceolate leaves on a short trunk, with erect branched axillary spikes of creamy white bell shaped flowers, the spike longer than the leaves.

TACCACEAE (175)

Tacca J.R. & G.Forst.

T. leontopetaloides (L.)O.Kuntze (*T. pinnatifida*)   'tavolo, tavoul, l'arouroute de france, indian arrowroot'

Summerhayes, Bailey, Lionnet (1), Wilson,(1), Robertson (6)

Localities recorded:M,P,An

Perennial herb with tuberous roots, solitary large trifoliolate pinnatifid leaf, the petiole to 1 m, and a solitary long-stalked cluster of green flowers and long thread-like bracts, followed by ovoid ribbed fruit to 3 cm wide. The tuber is edible and the stem fibre used to be plaited for hats etc.

DIOSCOREACEAE (176/1)

Dioscorea L.

D. alata L.   'cambare, greater, water, winged, asiatic or white yam'

Summerhayes, Bailey, Fosberg (1)

Rob 3151(S)

Localities recorded: M(sr only),Co,Coe

Cultivated climbing plant with square winged stems, twining to the right, heart-shaped leaves to 30 cm long, 3-winged fruit to 3.5 cm wide and large cylindrical tubers, sometimes lobed but usually single, to 10 kg or more, with edible flesh varying from white to deep purple. There may also be a few small aerial tubers.

D. bemarivensis Jum. & Perr.   'cambare'

Renvoize, Fosberg (2)

Localities recorded: Ald,Ass,Ast

Slender climber with small globose tubers with projections, 3- or 5-foliolate leaves, the leaflets to 10 cm long, and hanging clusters of 3 winged fruit, each to 1.5 cm wide.

D. bulbifera L.   'pomme edouard, morts aux rats, potato or aerial yam'

Summerhayes

Localities recorded: M

Cultivated plant with cylindrical spineless stem, twining to the left, with large heart-shaped leaves to 30 cm long, winged fruits to 2.5 cm long and edible aerial tubers, flattened and greyish brown, which can be up to 2 kg, but which may require detoxification. The root tubers are usually reduced and bitter or absent.

D. esculenta (Lour.)Burkhill (*D.aculeata*)   'cambare beti, lesser yam'

Summerhayes, Bailey

Localities recorded: M

Cultivated climbing plant with cylindrical spiny stem, twining to the left, with heart-shaped leaves and clusters of small ovoid tubers to 20 cm long with thin brownish skin and edible whitish flesh.

LILIACEAE (178/1)

Agapanthus L'Hérit.

A. africanus (L.)Hoffm.   'agapanthus'

Bailey

No locality recorded

Ornamental bulbous-rooted plant to 1 m tall with strap-shaped leaves and long-stalked heads of blue flowers, each one stalked and up to 4 cm long, and up to 30 in the cluster. There is a white variety.

Allium L.

A. cepa L. (incl. var. *aggregatum* G.Don and var. *cepa*)   'oignon, onion (var. *cepa*), shallot (var. *aggregatum* not Bailey's *A. ascalonium*)'

Bailey

No locality recorded

Cultivated herb with prominent bulb (var. *cepa*) or many well formed small bulbs (var. *aggregatum*), narrow hollow leaves to 40 cm long, almost round in cross-section, heads of greenish white flowers opening irregularly. The bulbs are eaten as vegetables.

A. fistulosum L.   'japanese bunching onion, spring onion'

Localities recorded: M(sr only)

Cultivated herb with little bulb development, but lateral buds growing to form a clump, the leaves to 50 cm long and hollow, round in cross section, the heads of yellow flowers opening regularly, from the apex downwards. The bulbs and leaves are eaten.

A. sativum L.   'ail, l'ail, garlic'

Bailey, Baumer

Cultivated herb to 60 cm with bulb made up of many small concave bulbs or cloves on a central stem within papery scales, and flat linear leaves, folded lengthwise. The bulbs are used as a flavouring and used medicinally in Seychelles.

A. schoenoprasum L.   'oignon plat (Bailey may have confused chives with chinese chives see below), chives'

Bailey

Localities recorded: M(sr only)

Cultivated herb to 30 cm with densely crowded small bulbs and clumps of narrow tubular leaves, bright green, to 25 cm long, tillering freely. The leaves are used as flavouring and garnish.

A. tuberosum Rottler ex Spreng.   'chinese chives, flat leaved garlic'

Localities recorded: M(sr only)

Cultivated rhizomatous herb with clumps of flat narrow leaves to 30 cm long, tasting of garlic and used as flavouring.

Asparagus L.

A. officinalis L.   'asperge, asparagus'

Bailey

Localities recorded: P(sr only)

Cultivated erect herb to 2 m with feathery branchlets and minute scale-like leaves with small axillary yellow flowers and globose red fruit. The young shoots are eaten as a vegetable.

A. setaceus (Kunth)Jessop (*A. plumosus*)   'fougere marriage, liane asperge, asparagus fern'

Bailey

Localities recorded: M(sr only)

Cultivated prickly climber with fern-like fronds of needle-like branchlets, small sessile flowers and globose fruit to 5 mm wide.

A. umbellulatus Bresler   'asperge'

Renvoize, Fosberg (2)

Localities recorded: Ald,Cos,Ast

Scrambling feathery shrub with scale-like leaves and leaf-like shoots with small cream flowers and pale orange brown globose fruits to 7 mm wide.

Chlorophytum Ker-Gawl.

C. comosum (Thunb.)Jacques   'spider plant'

Localities recorded: M(sr only)

Cultivated tuberous-rooted plant with rosettes of grass-like yellow and green striped leaves to 30 cm long and yellow long-stemmed spikes of small white flowers followed by plantlets which root freely.

Cordyline Comm, ex Juss.

C. fruticosa (L.)A.Chev.   '?chandelle rouge'

Robertson (7)

Localities recorded: M(sr only),Dar

Cultivated shrub to 4 m with stems showing prominent leaf scars, ovate lanceolate leaves (one variety with wider and softer coloured leaves), variegated in shades of pink, red, green and white, and axillary branched spikes of small creamy pink flowers.

Dianella Lam.

D. ensifolia (L.)DC.   'manga save'

Summerhayes, Bailey, Lionnet (1), Baumer

Ves 5583(K), Arc 129(K), Jef 362,371(S), Sch 11653(K), Lam 410(K), Fos 51967(S), Rob 2533(S)

Localities recorded: M

Rhizomatous perennial to 1m with linear leaves in two ranks and slender stems bearing small blue flowers and bright blue ovoid fruits to 1.5 cm wide. Used medicinally in Sehychelles.

Dracaena L.

D. reflexa Lam. var. **angustifolia** Baker, (*Pleomele angustifolia*)   'bois chandelle blanc, bois chandelle rouge (but Bailey probably refers to *Cordyline fruticosa*)'

Summerhayes, Bailey, Renvoize, Fosberg (2), Baumer, Wilson (1),(2)

Jef 705(S), Rob 2594,2670(S)

Localities recorded: M,P,F,N,C,L,Ald,Ass

Indigenous slender shrub to 5 m with lanceolate leaves to 30 cm long, clustered at the ends of the branches, and a branching spike of greenish white flowers, each to 1.5 cm long, and orange or red globose fleshy fruit to 1 cm wide. Used medicinally in Seychelles.

Gloriosa L.

G. superba L. (*G.rothschildiana*)   'gloriosa, glory lily'

Bailey, Robertson (6)

Localities recorded: M(sr only),An

Tuberous-rooted climber with lanceolate leaves, the tips prolonged into tendrils, and striking flowers with 6 recurved dark red, or red and orange, perianth segments, the stamens spreading and the style at right angles to the flower axis. (Fig. 175)

Lomatophyllum Willd.

L. aldabrense Marais (*Lomatophyllum borbonicum*)   'zanana mouro'

Renvoize, Fosberg (2)

Localities recorded: Ald,Ass,Ast

Endemic perennial with rosette of fleshy lanceolate leaves to 1m long on a stout stem, with a branched spike of reddish yellow flowers, each to 2.5 cm long, and purplish red fruit to 1.2 cm wide.

Ophiopogon Ker-Gawl.

O. intermedius D.Don

Rob 3183(K)

Localities recorded: M

Ornamental tuberous rooted perennial with dark green narrow strap-shaped leaves to 40 cm long, and stiff erect spikes bearing small white flowers, each to 5 mm long, in few flowered clusters, the flowers hanging downwards.

Sansevieria Thunb.

S. thyrsiflora Thunb. (*S.guineensis*)   'sanseveria'

Bailey, Lionnet (1)

No locality recorded

Rhizomatous plant with a few stiff erect broadly lanceolate leaves to 50 cm long, mottled dark and pale green, with terminal spikes of whitish flowers, each to 2 cm long, and globose orange fruit to 8 mm wide.

S. trifasciata Hort. ex Prain   'mother in law's tongue'

Localities recorded: M(sr only)

Ornamental plant similar to *S. thyrsiflora* but with leaves to 1m and narrowly lanceolate, the mottling white and green, or greenish with yellow edge to the leaves. (Fig. 176)

PONTEDERIACEAE (179)

Eichhornia Kunth

E. crassipes (Mart.)Solms-Laub.   'water hyacinth'

Bailey, Lionnet (2)

Localities recorded: M(BG)

Rhizomatous aquatic herb with feathery roots, round leaves with inflated petioles and erect spikes of showy blue flowers, each to 2.5 cm wide.

COMMELINACEAE (183/1)

Commelina L.

C. benghalensis L.   'herbe cochon, l'herbe cochon grand feuilles'

Summerhayes, Renvoize, Fosberg (2), Wilson (2),(4)

Bru/War 96(S), Rob 3141,2628(S)

Localities recorded: S,N,C,A,Coe,Mar,Ass

Semi-prostrate pubescent annual herb with succulent stems, rooting at the nodes, ovate leaves to 7 cm long and clusters of blue flowers, with two large and one small petal, enclosed in a bract, and an oblong capsule to 5 mm long. (Fig. 177)

C. diffusa Burm. f. (*C. nudiflora*)   'l'herbe cochon'

Summerhayes, Gwynne, Fosberg (1), Stoddart (1)

Jef 650 (S)

Localities recorded: S,N,Co,?Rem

Prostrate glabrous herb with oblong-lanceolate leaves to 7 cm long and clusters of flowers, with the large petals blue and the small ones white, enclosed in a stalked bract.

C. salicifolia Roxb.   'herbe cochon'

Summerhayes, Robertson (7)

Jef 734(S), Rob 2365(S)

Localities recorded: M,L,Dar,Poi

Trailing plant with lanceolate leaves to 8 cm long, slightly succulent stems and few small bright blue flowers from boat-shaped bracts.

Tradescantia

T. spathacea Swartz (*Rhoeo discolor, R. spathacea*)   'corbeille de moise, adam and eve, boat lily, oyster plant'

Bailey, Robertson (7)

Localities recorded: M(sr only),Dar

Ornamental erect succulent plant to 50 cm tall with lanceolate leaves to 25 cm long, bronze green above and purplish red below, with axillary cluster of white flowers held in paired purple boat-shaped bracts.

Zebrina Schnitzl.

Z. pendula Schnitzl.   'zebrina, cockroach grass, wandering jew'

Bailey

Localities recorded: M(sr only)

Ornamental creeping plant with fleshy stems, rooting at the nodes, and ovate greyish green and purplish striped leaves to 6 cm long, with small purplish red flowers with white stamens.

FLAGELLARIACEAE (185)

Flagellaria L.

F. indica L.   'rattan'

Summerhayes, Wilson (2)

Jef 619(S), Ber 14618(K), Rob 2709(S)

Localities recorded: M,S,P,C,Fe

Indigenous climber to 10 m with lanceolate leaves to 35 cm long, the tips becoming tendrils, with terminal sprays of small white flowers, each to 2 mm wide, and globose red fruit to 7 mm wide. (Not to be confused with the foliage of *Gloriosa superba* which it resembles) (Fig. 178)

JUNCACEAE (186/1)

Juncus L.

J. squarrosus L.   'herbe la mar, refle'

Bailey

No locality recorded

Densely tufted perennial with basal rosette of rigid linear leaves and erect flowering stems to 50 cm with terminal clusters of dark brown florets each with 6 perianth segments. (This seems unlikely in Seychlelles and may have been a misidentification for a sedge, perhaps *Fimbristylis complanata*)

PALMAE (187)

Acrocomia Mart.

A. aculeata (Jacq.)Lodd. ex Mart.   'gru gru palm'

Lionnet (2)

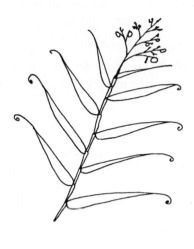

Fig. 177. Commelina benghalensis   Fig. 178. Flagellaria indica

Fig. 179. Areca catechu   Fig. 180. Corypha umbraculifera

Localities recorded: M(BG)

Cultivated palm to 20 m with trunk thickened towards the top with rings of spines, pinnate leaves to 3 m long, finely hairy beneath, ♂ and ♀ flowers together in a hairy inflorescence. The fruits globose to 3.5 cm wide.

Areca L.

A. catechu L.   'arequier, coco pacques, betel palm'

Bailey, Lionnet (2), Wilson (1), Robertson (4)

Localities recorded: M(BG),P,F

Cultivated slender palm to 30 m with ringed trunk, pinnate leaves to 1.5 m long and a crownshaft below which emerge the branched inflorescences, ♂ and ♀ flowers together, and yellow to orange ovoid fruits to 10 cm long. The seeds are chewed with *Piper betel* as a masticatory. (Fig. 179)

Borassus L.

B. flabellifer L.   'palmyra'

Bailey, Lionnet (2)

Jef 1165(K)

Localities recorded: M(BG)

Cultivated palm to 25 m with palmately divided leaves to 4 m long and sometimes a slight swelling above the middle of the trunk, ♂ and ♀ flowers on separate trees and subglobose edible orange fruit to 15 cm wide. Fibre is obtained from the leaf bases and the flower shoots are tapped for toddy.

Brahea Mart. ex Endl.

B. edulis Wendl. ex Watson (*Erythea edulis*)   'guadalupe palm'

Lionnet (2)

Localities recorded: M(BG)

Cultivated palm with stout trunk, to 10 m or more with deeply palmately divided leaves to 2 m wide with stout prickly petioles, the flowers bisexual and the globose black fruit to 2.5 cm wide.

Caryota L.

C. mitis Lour.   'clustered fishtail palm'

Lionnet (2)

Localities recorded: M(BG)

Cultivated clustered palm to 10 m, with bipinnate leaves, the pinnules wedge-shaped, and densely branched inflorscences of ♂ and ♀ flowers together, with red fruit to 1 cm wide. After flowering that stem dies.

C. urens L.   'toddy palm, fish tailed palm, indian wine palm'

Bailey, Lionnet (2)

Localities recorded: M(BG)

Cultivated solitary palm to 20 m with bipinnate leaves to 6 m, the pinnules wedge-shaped, inflorescence of ♂ and ♀ flowers together and globose fruit to 2 cm wide. After flowering the tree dies. The inflorescences are tapped for toddy and sugar and fibre can be obtained from the leaves.

Chrysalidocarpus Wendl.

C. lutescens Wendl.   'cane palm, golden feather palm'

Lionnet (2)

Localities recorded: M(BG)

Cultivated clustered palm to 10 m or more with erect pinnate leaves, curved at the top, the pinnae to 60 pairs and in one plane, ♂ and ♀ flowers together on the inflorescence and oblong violet black fruit to 2 cm long.

C. madagascariensis Becc. (*C. lucubensis*)

Lionnet (2a)

Localities recorded: M(BG)

Cultivated clustered or solitary palm to 10 m with pinnate leaves with up to 100 pairs of pinnae, in groups and in several planes, with rusty tomentose inflorescences, ♂ and ♀ flowers together, and ovoid fruit to 1.7 cm long.

Cocos L.

C. nucifera L.   'cocotier, coconut palm'

Summerhayes, Sumner Gibson, Bailey, Piggott, Gwynne, Sauer, Lionnet (1),(2), Fosberg (1),(2), Stoddart (1),(2),(3), Baumer, Wilson (1),(3),(4), Robertson (1),(2),(3),(4),(5),(6),(8)

Localities recorded: M,S,P,F,N,C,An,Co,V,Bir,Den,Pla,Coe,Afr,Rem,Des, Dar,Jos, Poi,Mar,Alp,Def,Pro,Far,Ald,Ass,Cos,Ast (also, as sr only,A,D,Ma,Cn, T,Ce,L,Cu, Mo,Ro,Ao)

Indigenous, and now widely cultivated palm to 30 m with pinnate leaves to 6 m with sheathing leaf bases, branched inflorescences with ♂ flowers at the top and ♀ flowers at the base, followed by the large ovoid 3-sided fruit to 30 cm long. The white endosperm of the nut is eaten and yields oil and all parts of the tree are used in Seychelles, and it is used medicinally. There is some debate about whether it is indigenous or was introduced.

Corypha L.

C. umbraculifera L.   'talipot palm'

Bailey, Lionnet (2)

Localities recorded: M(BG)

247

Cultivated palm to 25 m with thick trunk covered with old leaf bases, round palmately divided leaves to 5 m wide and a vast terminal pyramidal inflorescence, produced after 30 to 40 years, with 3 cm wide fruits which take a year to ripen. After flowering the tree dies. (Fig. 180)

C. utan Lam. (*C. elata*) 'gebang palm'

Lionnet (2a)

Localities recorded: M(BG)

Cultivated palm to 20 m with wide ringed trunk with a spiral furrow, palmate leaves to 2 m or more wide, with up to 100 segments, conical terminal inflorescences of bisexual flowers to 3 m long and globose fruit to 2.2 cm wide. After flowering the tree dies.

Cyrtostachys Blume

C. renda Blume (*C. lakka*) 'sealing wax palm'

Bailey, Lionnet (2), Procter (2)

Localities recorded: M(BG),P

Cultivated cluster palm to 10 m tall, with pinnate leaves to 2 m, and sheaths and petioles bright orange red, with ovoid conical fruit to 2 cm long.

Daemonorops Blume

D. ?jenkinsiana (Griff.)Mart. 'rotin, rattan'

Bailey, Lionnet (2) (not *Calamus sp.*)

Localities recorded: M(BG)

Cultivated climbing palm with various and numerous spines and thorns, with the ends of the pinnate leaves with long whip-like projections.

Deckenia Wendl.

D. nobilis Wendl. 'chou palmiste, palmiste, millionaires salad'

Summerhayes, Bailey, Lionnet (1),(2), Procter (1),(2)

Jef 560,775,1227(K)

Localities recorded: M,S,P,D,C,Fe

Endemic palm to 40 m with pinnate leaves to 5 m long and a smooth crownshaft below which are the inflorescences of long drooping yellow branches of ♂ and ♀ flowers, emerging from a prickly spathe held at right angles to the trunk, and ovoid 1 cm long purplish black fruits. There may be more than one species involved.

Dictyosperma Wendl. & Drude

D. album (Bory)Balf.f.

Summerhayes, Lionnet (2)

Localities recorded: M(BG)

Cultivated palm to 20 m with pinnate leaves to 4 m long, with a crownshaft and branched inflorescences of ♂ and ♀ flowers and purplish olive-like fruit to 1 cm long. (Fig. 181)

Elaeis Jacq.

E. guineensis Jacq.   'palm oil palm'

Bailey, Lionnet (2)

Localities recorded: M(BG)

Cultivated and escaping erect palm to 20 m with persistent leaf bases for several years, pinnate leaves to 4.5 m long, each pinna an inverted V in cross-section, and the lower pinnae reduced to spines, ♂ and ♀ flowers in separate inflorescences, the fruits borne in large dense masses, each fruit ovoid to 2 to 5 cm long, the commonest form whitish below and voilet to black above. The oil is expressed from the fruit. (Fig. 182)

Hyophorbe Gaertn.

H. lagenicaulis (L.H.Bail.)Moore (*Mascarena revaughnii*)   'round island palm, bottle palm'

Lionnet (2)

Localities recorded: M(BG)

Cultivated solitary palm with trunk swollen at the base, pinnate leaves above a crownshaft, below which is the branched inflorescence of ♂ and ♀ flowers in groups, followed by the purplish ovoid fruit to 2.5cm long containing one seed.

Jubaea Humboldt

J. chilensis Molina (*J. spectabilis*)   'coquito palm, chilean wine palm'

Lionnet (2)

Localities recorded: M(BG)

Cultivated solitary palm to 10 m or more with stout trunk, pinnate leaves to 2 m, curved at the top, ♂ and ♀ flowers in the same inflorescence, maroon, with ovoid yellow fruit to 3 cm long.

Latania Comm. ex Juss.

L. loddigesii Mart.   'blue latan'

Lionnet (2)

Localities recorded: M(BG)

Cultivated palm to 20 m with glaucous long-petioled deeply palmately divided leaves to 2 m long, ♂ and ♀ flowers on separate trees with fruit to 5 cm long which is ridged at the base and ribbed at the apex.

L. lontaroides Gaertn. (*L. commersonii, L. borbonica*)   'red latan'

Summerhayes, Lionnet (2), Robertson(5)

Localities recorded: M(BG),N

Cultivated palm to 15 m with long-petioled deeply palmately divided leaves, greyish green but reddish when young, to 2 m long, ♂ and ♀ flowers on separate trees and globose fruits to 4 cm wide.

L. verschaffeltii Lemaire 'yellow rattan'

Lionnet (2)

Localities recorded: M(BG)

Cultivated palm to 20 m with deeply palmately divided leaves, pale green and edged yellow, to 1.5 m long, and very long inflorescences with ovoid fruit to 4.5 cm long, flat-sided and bilobed at the apex.

Licuala Thunb.

L. spinosa Thunb.

Lionnet (2)

Localities recorded: M(BG)

Cultivated clustered palm to 5 m with palmately divided leaves to 1.5 m wide, with blunt-ended segments, and a 2 m long inflorescence of bisexual flowers with orange red globose fruit to 1.5 cm long.

Livistona R.Br.

L. chinensis (Jacq)R.Br.

Lionnet (2a) (not *L. decipiens*)

Localities recorded: M(BG)

Cultivated palm to 15 m with palmately divided leaves to 2 m long, the segments drooping at the apex, bisexual flowers in small groups on the branched inflorescence and blackish subglobose fruit to 1.5 cm long.

L. rotundifolia (Lam.)Mart.

Lionnet (2)

Jef 1166,1167(K)

Localities recorded: M(BG)

Cultivated palm to 20 m with ringed trunk, round palmate leaves with segments bifid at the tips, to 1 m wide, spiny petioles and hanging inflorescence with yellowish globose fruit to 1.5 cm wide.

Lodoicea Comm. ex Labill.

L. maldivica (Gmel.)Pers. 'coco de mer, double coconut'

Summerhayes, Bailey, Lionnet (1),(2), Procter (1),(2)

Fig. 181. Dictyosperma album   Fig. 182. Elaeis guineensis

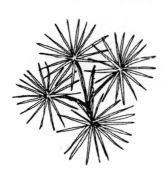

Fig. 183. Sabal minor   Fig. 184. Pandanus sechellarum

251

Jef 1166,1234(K)

Localities recorded: P,C (cultivated on M,S,D,F,Fe)

Endemic palm to 30 m with large rigid fan like leaves to 7 m with petioles to 4 m, ♂ and ♀ flowers on separate trees, the ♂ flowers in rigid catkins to 2m long and the ♀ flowers on 70 cm long zigzag spikes, followed by the large olive normally bilobed fruit to 40 cm long.

Nephrosperma Balf.f.

N. vanhoutteanum (Wendl. ex van-Houtt.)Balf.   'latanier millepatte'

Summerhayes, Lionnet (1),(2), Procter (1),(2), Robertson (6)

Jef 408(S),784,1218,1236,1238(K), Rob 2464(S)

Localities recorded: M,S,P,C,An

Endemic palm to 14 m with pinnate leaves to 3 m with mealy leaf bases, stiff branched inflorescences of ♂ and ♀ flowers together, and flattened globose orange red fruit to 1.5 cm wide.

Orbignya Mart. ex Endl.

O. cohune (Mart.)Dahlgren (*Attalea cohune*)   'cohune palm'

Bailey

No locality recorded

Cultivated palm with long feathery pinnate leaves to 9 m long, with egg-shaped brown fruit to 8 cm long in large clusters. The nuts yield cohune oil.

Phoenicophorium Wendl.

P. borsigianum (K.Koch)Stuntz (*P. sechellarum*)   'latanier feuille'

Summerhayes, Bailey, Lionnet (1),(2), Procter (1),(2), Robertson (6)

Jef 471,643(S),558,565,1221(K)

Localities recorded: M,S,P,D,F,C,Fe,An

Endemic palm to 20 m with spiny stem when young, leaves to 1.5 m long split at the apex only and with bifid orange edged serrations, ♂ and ♀ flowers together on a stiff branched inflorescence followed by small ovoid orange red fruit to 1.5 cm long.

Phoenix L.

P. dactylifera L.   'dattier, datt, date palm'

Bailey, Fosberg (2)

Localities recorded: M(sr only),Ald

Cultivated and escaping palm to 30 m with thick trunk, persistent leaf bases, pinnate leaves to 7 m long, the pinnae in groups and each pinna V-shaped in section, and the lower pinnae reduced to spines, ♂ flowers on one tree, ♀ flowers on another, the fruits oblong, yellow to brown or black to 7cm long with edible flesh.

P. loureirii Kunth (*P. humilis*)

Lionnet (2)

Localities recorded: M(BG)

Cultivated, often clustered, palm, usually without trunk, with pinnate leaves to 2 m, the pinnae in groups and V-shaped in section, ♂ flowers on one tree, ♀ on another, with red or purple ovoid fruits to 1.5 cm long.

P. rupicola T.Anders.

Lionnet (2a)

Localities recorded: M(BG)

Cultivated slender palm to 7 m with pinnate leaves to 3 m, the blade twisted but the pinnae all in one plane, and oblong glossy yellow fruit to 1.7 cm long.

Ptychosperma Labill.

P. macarthurii (Wendl.)Nichols. (*Actinophloeus macarthurii*) 'tire couyon, hurricane or macarthur palm'

Bailey

Jef 593(K)(not *A. hospitum*)

Localities recorded: M(BG)

Cultivated slender palm to 7 m or more with long pinnate leaves to 2 m, the pinnae blunt and wide, and red ovoid fruit to 2cm long hanging in dense clusters. *P. elegans* Blume, a single stemmed palm, has also been seen.

Raphia P. Beauv.

R. farinifera (Gaertn.)Hylander (*R. ruffia*)   'raffia'

Summerhayes, Bailey, Lionnet (1),(2), Wilson (1)

Localities recorded: M,P

Cultivated and naturalized palm to 15 m with large feathery pinnate leaves to 12 m long and terminal inflorescences of hanging clusters of shiny brown cone-like fruit, each to 5 cm long, hidden in the bracts. After flowering that stem dies. The leaves are used for weaving etc.

Rhapis L.f.

R. excelsa (Thunb.)Henry   'broad leaf lady palm, fern rhapis, dwarf ground rattan, bamboo palm'

Lionnet (2)

Localities recorded: M(BG)

Cultivated clustered palm to 4m, the trunks covered in coarse fibre, with palmately divided leaves with drooping and pleated segments, ♂ and ♀ flowers on separate trees, the fruit small fleshy and globose and containing one seed.

Roscheria Wendl.

R. melanochaetes (Wendl.)Wendl. ex Balf.   'latanier hauban'

Summerhayes, Bailey, Lionnet (2), Procter (1)

Jef 728(S), 465,573(K), Rob 2639(S)

Localities recorded: M,S,P

Endemic palm to 8 m with slender trunk, ringed with black spines, and irregularly pinnate leaves to 2.5 m long, the young leaves red, and a branched inflorescence of ♂ and ♀ flowers with small black elliptical fruit to 5 mm long.

Roystonea O.F.Cook

R. regia (Kunth)O.F.Cook (*R. elata, Oreodoxa regia*)   'royal palm'

Bailey, Lionnet (2)

Localities recorded: M(BG)

Cultivated palm to 25 m with smooth trunk, thickened about the middle, plume-like pinnate leaves with a crownshaft of smooth leaf bases and small globose purplish fleshy fruit. (*R. regia* is the commonly cultivated Cuban species, *R. elata* is a wild Florida species and unlikely to have been recognized by Bailey)

Sabal Adans.

S. minor (Jacq.)Pers.   'bush palmetto, dwarf palmetto'

Lionnet (2)

Localities recorded: M(BG)

Cultivated palm with short or no trunk, palmately divided glaucous leaves and erect inflorescences of whitish bisexual flowers with black glossy globose fruit to 1.5cm wide containing one slightly flattened seed. (Fig. 183). *S. mauritiiformis* Becc. has also been seen.

Syagrus Martius

S. romanzzoffiana (Cham.)Glassman (*Arecastrum romanzoffianum*) 'queen palm'

Lionnet (2)

Localities recorded: M(BG)

Cultivated palm to 15 m or more with arching pinnate leaves to 5 m, the pinnae in groups and in several planes, ♂ and ♀ flowers in the same inflorescence, to 1 m long, with yellow ovoid fruit to 2.5 cm long.

Thrinax Sw.

T. parviflora Sw.   'thatch palm'

Lionnet (2a)

Localities recorded: M(BG)

Cultivated slender palm to 10 m or more with palmately divided leaves to 1 m long, bisexual flowers and stalked globose white fruit to 5 mm wide. (This name is possibly incorrect)

Verschaffeltia Wendl.

V. splendida Wendl.   'latanier latte, latte'

Summerhayes, Bailey, Lionnet (1),(2), Procter (1),(2)

Jef 1235(K), Rob 2463(S)

Localities recorded: M,S,P

Endemic palm to 30 m with ringed trunk and cone of stilt roots at the base, simple pinnately veined leaves to 3 m long, the margins regularly toothed, ♂ and ♀ flowers together on drooping inflorescences and globose red fruit to 2 cm wide.

PANDANACEAE

Pandanus Parkinson

P. aldabraensis St. John (*P. vandermeeschii*)

Renvoize, Lionnet (2), Fosberg (2)

Localities recorded: Ald

Endemic small tree to 2.5 m with linear leaves to 60 cm long, weakly prickly, and cone-like fruit with the top of each section showing one apex.

P. balfourii Mart.   'vacoa de riviere, balfour's pandanus'

Summerhayes, Lionnet (1), Procter (1)

Jef 631,692(S), Fos 52092(S), Bru/War 89(S), Rob 2551, 2622(S)

Localities recorded: M,S,P,D,F,N,C,A,Fe,Co,Cu

Endemic slender tree to 10 m, or shrub with mass of old leaves, drooping leaves to 2 m with white marginal spines, and oblong fruits to 25 cm long, the sections with variably grooved top with several apices. (This species may be synonomous with *P. odoratissimus* Balf.)

P. hornei Balf.f.   'vacoa parasol, horne's pandanus'

Summerhayes, Bailey, Lionnet (1), (2), Procter (1),(2)

Jef 460(S)

Localities recorded: M,S,P,C

Endemic large tree, usually branching in 3's, to 20 m tall, the crown umbrella-shaped, with stilt roots at the base of the trunk, linear leaves with red marginal spines, to 3.5 m long and cone-like fruits to 35 cm wide, the top of each section with ridges.

P. multispicatus Balf.f.   'vacoa de montagne, vacoa millepatte'

Summerhayes, Bailey, Procter (1),(2), Robertson (4)

Ves 5478(K), Jef 557,563(S)

Localities recorded: M,S,P,F,C

Endemic shrub to 4 m, sometimes spreading, with very prickly linear leaves to 1.5 m long, with oblong cone-like fruit to 4 cm long, the individual sections to 5 mm long, several together at the top of the plant.

P. sanderi Hort.   'variegated screwpine'

Lionnet (2), Robertson (7)

Localities recorded: M(BG),Dar

Cultivated screwpine with short stem, suckering freely, and long spiny margined yellow and green striped variegated leaves to 2 m, bent in the middle when old.

P. sechellarum Balf.f.   'vacoa marron, seychelles pandanus'

Summerhayes, Bailey, Lionnet (2), Procter (1),(2)

Jef 1224(S),460(K), Ber 14656(K)

Localities recorded; M,S,P,D

Endemic tree to 15 m with stilt roots from the leaf clusters at the ends of the branches, linear leaves to 3 m long with pinkish spines and stalked cone-like fruit to 30 cm wide, each section deeply grooved on top giving prominent lobes. (Fig. 184)

P. tectorius Park.   'vacoa, vacquois'

Renvoize, Fosberg (2)

Localities recorded: Ald,?Cos

Sparsely branched tree with well developed stilt roots, prickly linear leaves to 1.5 m and hanging cone-like fruit to 25 cm wide with many angular sections each with a varying number of apical lobes. (This may also be synonomous with *P. ordoratissimus*)

P. utilis Bory   'vaccoa corde, vacoa de pays, vacoa sac, useful pandanus'

Bailey, Gwynne, Lionnet (2), Procter (2), Robertson (4)

Rob 2619,3108,3164(S)

Localities recorded: M(BG),P,F,N,Coe

Cultivated and escaping stout branching tree to 15 m with stilt roots and erect slightly glaucous leaves to 2 m long with red marginal spines, and large cone-shaped fruit to 15 cm wide hanging down on long stalks.

TYPHACEAE (190/1)

Typha L.

T. javanica Schnitzl. ex Zoll. (*T. angustifolia*)   'herbe zonc, jonc, bullrush, cattail'

Summerhayes, Bailey, Stoddart (3), Wilson (2), Robertson (5)

Jef 1209(S), Rob 2801(S)

Localities recorded: S,P,N,C,?Bir,Den

Indigenous perennial erect marsh plant to 2 m with linear leaves and cylindrical brown flower spikes, on stems to 2 m, the densely arranged ♀ flowers forming a 2 cm wide column with the ♂ flowers above.

ARACEAE (191)

Alocasia (Schott)G.Don

A. macrorrhiza (L.).G.Don   'tannia, via'

Summerhayes, Bailey, Gwynne, Lionnet (2), Fosberg (1) Stoddart (1),(2),(3), Wilson (1),(3),(4), Robertson (2),(3),(4)

Bur/War 97(S), Rob 2889(S)

Localities recorded: ?M(BG),P,F,A,Bir,Den,Coe,?Des,Dar,?Poi,Mar,Def,Alp

Cultivated large erect herb to 4 m tall with stout stem, large triangular leaves with deep basal lobes to 1m long, and boat-shaped spathes with a spadix to 25 cm long carrying the globose red fruits. The stem is the edible portion but contains calcium oxalate raphides which have to be removed by cooking.

Amorphophallus Blume

A. paeoniifolius (Dennst.)Nicolson (*A. campanulatus*)   'elephant yam, saoule mouche, giant aroid'

Bailey, Lionnet (1),(2)

Bogner sn(K)

Localities recorded: M(BG),S,P

Plant with large hemispherical tuber to 25 cm wide, depressed in the centre, and a solitary tripartite pinnatifid leaf to 1m wide which dies away and is followed by a large foetid sessile spathe to 30 cm wide with undulating margins and a mushroom-shaped spadix. The tuber is edible but contains calcium oxalate crystals which have to be removed by prolonged boiling.

Anthurium Schott

A. spp. (incl. *A. andraeanum* Linden and *A. crystallinum* Linden)   'anthurium, anthurium rouge'

Bailey

Localities recorded: M(sr only)

257

Ornamental plants with large arrow-shaped leaves, sometimes grown for the foliage (*A. crystallinum* has velvety green leaves with white veins) but more often for the spectacular flowers with coloured flat spathes and erect spadices as in *A. andraeanum* and its various hybrids and cultivars.

Caladium Vent.

C. spp. (incl. *C. bicolor* (Dryand.)Vent. and *C. argyrites* Lem.)   'caladium grand feuille (*C. bicolor*), caladium petit feuille (*C. argyrites*)'

Bailey, Robertson (4)

Localities recorded: M(sr only),F

Ornamental and escaping tuberous-rooted plants with large peltate, heart-shaped or arrow-shaped leaves with variously coloured, usually red, pink or white, blotches and veins, and solitary flowers with spathe inflated at the base and widening to show the erect spadix inside.

Colocasia Schott

C. esculenta (L.)Schott (incl. var. *esculenta* and var. *antiquorum*)   'arouille, songe blanc, taro, cocoyam, dasheen, edoe'

Summerhayes, Bailey, Gwynne, Stoddart (3), Robertson (2),(4)

Fos 52115(S), Bru/War 98(S)

Localities recorded: M(sr only),F,A,Co,Bir,Coe

Cultivated herb with underground edible starchy corm (large and cylindrical and to 30 cm long in var. *esculenta*, smaller and with many side tubers in var. *antiquorum*), triangular peltate leaves with rounded basal lobes to 50 cm long and petioles to 1 m long. The spadix protrudes from the spathes in var. *esculenta*, but not in var. *antiquorum*. There is much confusion over the common names.

Dieffenbachia Schott

D. sequine (Jacq.)Schott (*D. maculata*)   'bois tangue, via tangue, dumb cane'

Bailey, Lionnet (1), Procter (2), Robertson (7)

Localities recorded: M,P,Dar

Ornamental plants with erect leafy stems to 2 m with leaf scars, ovate green and white blotched leaves to 30 cm long and poisonous sap which irritates the mucous membranes and causes massive swelling of the mouth and throat and can choke and stop speech.

Monstera Adans.

M. deliciosa Liebm.   'taro vine, monstera'

Lionnet (1), Wilson (1), Robertson (7)

Localities recorded: M,P,Dar

Cultivated woody climber with hanging cord like roots, large pinnately lobed and perforated leaves and flowers of open spathe and columnar spadix followed by a cone-like fruit, the stylar part of the ♀ flowers fused together and peeling off to reveal the seeds in edible orange pulp.

Philodendron Schott

P. spp. (incl. *P. bipinnatifidum* Schott and *P. scandens* C.Koch & H.Sello) 'philodendron'

Lionnet (1), Procter (2), Robertson(7)

Localities recorded: M,P,Dar

Vigorous ornamental and widely escaping climbing woody plants with handsome variously shaped glossy leaves and roots clinging to the trees they are climbing on.

Protarum Engl.

P. sechellarum Engl.   'arouroute de linde marron'

Summerhayes, Lionnet (1), Procter (1)

Jef 485(S), Sch 11677(K), Bogner 223(K)

Localities recorded: M,S,P

Endemic tuberous-rooted plant with solitary deeply palmately lobed leaf to 30 cm wide, with up to 8 lobes, the solitary flower arising from the root with a simple spathe and columnar spadix.

Syngonium Schott

S. ?podophyllum Schott

Localities recorded: M(sr only)

Ornamental climber with clasping rootlets, milky sap, digitately compound leaves to 25 cm long, the leaflets sometimes lobed, but leaves arrow-shaped when young, with several spathes per node, the spathe creamy white to 10 cm long, inflated below and widening above, the upper part withering in fruit and the remainder, containing the fruiting spadix, deflexed. (Fig. 185)

LEMNACEAE (192)

Lemna L.

L. sp.   'duckweed'

Localities recorded: M(sr only)

Aquatic plant consisting of one ovate 'leaf' to 4 mm long, floating on the surface of open water in marshes.

TRIURIDACEAE (193)

Seychellaria Hemsley

S. thomassetii Hemsley

Summerhayes, Procter (1)

Jef 579(K)

Localities recorded: M

Endemic saprophytic (? on rotting *Pandanus sechellarum* leaves) plant with purplish stems and tiny flowers, drying yellow, to about 10 cm tall, with colourless underground parts. Probably very ephemeral.

NAJADACEAE (195/1)

Najas L.

N. australis Bory ex Rendle (*N. setacea*)

Summerhayes

No locality recorded

Aquatic plant with many branched thread-like green stems and whorls of linear serrated leaves, each to 1 mm wide and 4 mm long.

N. graminea Delile

Fosberg (2)

Localities recorded: Ald

Plant of brackish water with slender stems and linear leaves to 2.5cm long and less than 1mm wide.

POTAMOGETONACEAE (195/6)

Potamogeton L.

P. thunbergii Chern. & Schltr. (*P. richardii*, ?*P. nodosus*)

Summerhayes

Jef 1169(S)

Localities recorded: M,P

Indigenous aquatic plant with long stems, narrow submerged leaves and ovate floating leaves to 12 cm long and 5 cm wide, erect dense cylindrical spikes of small flowers, the spikes to 6 mm wide and 5 cm long.

RUPPIACEAE (195/7)

Ruppia L.

R. maritima L.

Fosberg (2)

Fig. 185. Syngonium podophyllum   Fig. 186. Bulbostylis barbata

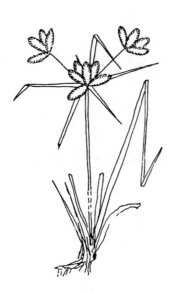

Fig. 187. Cyperus compressus   Fig. 188. Cyperus rotundus

Localities recorded: Ald

Brackish water or marine plant with long weak slender branching stems and linear thread-like leaves to 10 cm long.

ZANNICHELLIACEAE (195/9)

Cymodocea Koenig

C. rotundata Ehrenb. & Hempr.

Renvoize, Fosberg (2)

Localities recorded: Ald

Marine 'grass' with slender rhizomes and erect leafy branches, the 'stems' to 5 cm long, with narrow flat leaves to 40 cm long and to 4 mm wide, with pale purplish sheaths to 10 cm long.

C. serrulata Aschers. & Magn.

Summerhayes, Fosberg (2)

Localities recorded: P,Ald

Marine 'grass' with slender rhizomes and erect leafy branches, 'stems' to 10 cm long with narrow flat leaves to 15 cm long and 5 mm wide, and sheaths to 3 cm long.

Halodule Endl.

H. uninervis (Forssk.)Aschers.

Summerhayes (*Zostera* sp.), Gwynne, Renvoize, Fosberg (2)

Localities recorded: P,Coe,Ald,Cos

Marine 'grass' with thin rhizomes and thread-like linear leaves to 3.5 mm wide with 2 lateral projections and a central one at the leaf tip.

H. wrightii Aschers.

Renvoize, Fosberg (2)

Localities recorded: Ald,Cos

Marine 'grass' with thin rhizomes and thread-like linear leaves to 1 mm wide with 2 lateral projections only at the leaf tip.

Syringodium Kutz.

S. isoetifolium (Aschers.)Dandy (*Cymodocea isoetifolia*)

Summerhayes, Renvoize, Gwynne, Fosberg (2), Stoddart (2)

Pro/Fra 4360(K)

Localities recorded: P,Dar,Ald

Marine 'grass' with slender rhizomes and erect leaves, round in cross-section, with sheath to 4 cm long and blade to 30 cm long.

Thalassodendron den Hartog

T. ciliatum (Forssk.)den Hartog (*Cymodocea ciliata*)

Summerhayes, Gwynne, Renvoize, Fosberg (2), Stoddart (1),(2),(3)

Pro/Fra 4362(K)

Localities recorded: P,Bir,Den,Rem,Des,Dar,Jos,Far,Ald,Cos

Marine 'grass' with branching rhizomes and long leafy branches, to 30 cm or more long, bearing broadly linear leaves, sometimes curved, to 15 cm long and 1.3 cm wide, with a pink to purple sheath to 3 cm long, widely open at the top, shed with the leaves.

CYPERACEAE (199)

Bulbostylis Kunth

B. barbata (Rottb.)C.B.Cl. (*Stenophyllus barbatus*)

Summerhayes

Fos 52060,52140(S), Pro 4271(S),4441(K)

Localities recorded: M,C,Co

Slender tufted sedge to 20 cm with tight heads to 1 cm wide of small brown florets. (Fig. 186)

B. basalis Fosberg

Fosberg (2)

Localities recorded: Ald

Endemic tiny tufted annual sedge to 7.5 cm tall with slender leaf-like floral bracts similar to the leaves.

B. hispidula (Thunb.)Svenson (*Bulbostylis hirta, Fimbristylis hispidula*)

Fosberg (2)

Localities recorded: Ass

Densely tufted sedge to 30 cm tall with compound inflorescence of single chestnut-brown spikelets to 1 cm long.

Cyperus L.

C. alopecuroides Rottb. all *Cyperus* spp. are called 'herbe oignon, laiche'

Horne 221(K)(not in Summerhayes), Pro 4257(S), Bru/War 67(K), War 186(S)

Localities recorded: M,A,Co

Stout perennial sedge to 2 m tall usually in standing water, with long leaves, silvery beneath, to 1.5 m long and a large inflorescence of many dull brown spikelets, each to 8 mm long, in oblong spikes to 4 cm long.

C. articulatus L.

Summerhayes

Jef 660(S), Pro 4373(S),4433(K), Rob 2609,2797(S)

Localities recorded: M,S,P,D,N

Indigenous leafless perennial sedge to 2 m, usually in water, with spreading compound inflorescences of shiny mid brown spikelets up to 4 cm long and joints showing in the stem.

C. bigibbosus Fosberg

Fosberg (2)

Localities recorded: Ald

Endemic loosely tufted perennial sedge to 40 cm tall with small globose heads of white spikelets which dry to brown. (Similar to *Kyllinga alba* and as it has two stigmas perhaps should be in *Kyllinga*)

C. bulbosus Vahl

Fosberg (2)

Localities recorded: Ass

Slender sedge to 20 cm tall from blackish bulbs with reddish brown or greenish spikelets, each to 1.5 cm long, on simple, or branched only at the base, spikes.

C. compressus L.

Summerhayes, Robertson (7)

Pro 4106,4117,4213(S), Bru/War 20(S), Rob 2627,2729(S)

Localities recorded: M,F,N,A,Dar

Tufted annual sedge to 50 cm tall with flattened oblong spikelets to 2 cm long, often golden brown edged with green. (Fig. 187)

C. conglomeratus Rottb. (incl. var. *effusus* (Rottb.) Kukenth. f. *pachyrrhizus* (Boeck.)Kukenth.) (*C. pachyrrhizus*)

Summerhayes, Bailey, Renvoize, Fosberg (2), Stoddart (3)

Ves 5650(K), Jef 1205(S), Pro 4037,4255(S)

Localities recorded: Cu,Bir,Den,Ald

Indigenous stiff densely tufted sedge to 50 cm or more tall, with many dull white or pale brown spikelets, each to 1.5 cm long, in compact globose heads to 3.5 cm wide.

C. difformis L.

Rob 2819(S)

Localities recorded: M

Erect perennial sedge to 50 cm tall in wet places, with small green spikelets, each to 8 mm long, congested into globose heads to 1.2 cm wide, several heads in each inflorescence, some sessile, some stalked.

C. exaltatus Retz.

Bailey

Jef 540(S)

Localities recorded: M

Stout tufted perennial sedge to 1.5 m tall, with a large compound inflorescence of cylindrical spikes to 5 cm long, composed of small rounded golden spikelets to 4 mm long.

C. halpan L. (C. haspan)

Pro 4263(S),4493,4666(K), Rob 2796(S)

Localities recorded: M,P

Slender perennial sedge to 50 cm tall, in wet places, with a branched inflorescence, with a conspicuous wide bract at the base, of solitary, or a few in a cluster, slender flattened green or golden spikelets to 1 cm long. The inflorescence can bend over and take root.

C. maculatus Boeck.

Rob 3086(S)

Localities recorded: Coe

Rhizomatous perennial sedge with a compound inflorescence of cylindrical upwardly curving spikelets to 2 cm long with pointed ends, pale brown.

C. niveus Retz. var. leucocephalus (Kunth)Fosberg (C. obtusiflorus)

Renvoize, Fosberg (2)

Localities recorded: Ald

Tufted perennial sedge to 50 cm tall with the inflorescence a dense head of white to pale brown flattened spikelets to 1.5 cm long.

C. papyrus L.   'papyrus'

Localities recorded: M(M.Beach Hotel pond)(sr only)

Ornamental perennial aquatic sedge to 2 m tall with large compound head of long stalked clusters of red brown florets, each to 1 cm long.

C. rotundus L. (incl. ssp. tuberosus (Rottb.)Kukenth. var. platystachys C.B.Cl.)

Summerhayes, Bailey, Gwynne, Stoddart (3)

Fra 422(S), Pro 4448(K), Rob 2264,2723,2813,2879,3104(S), 2807(K)

Localities recorded: M,F,Den,Coe,Alp

Variable perennial sedge, a weed, with slender tuber-bearing stolons, to 75 cm tall with a compound inflorescence of long compressed spikelets, chestnut brown with golden edges, to 2 cm long (pale brown and green and to 3 cm long in ssp.) (Fig. 188)

Eleocharis R.Br.

E. dulcis (Burm.f.)Trin. (*E. plantaginoidea*)

Summerhayes

Jef 539(S), Pro 4142,4270(S),4389(K), Rob 2540(S)

Localities recorded: M,S,C

Indigenous leafless rhizomatous sedge growing in water with erect bright green stems to 1 m tall, showing the septa, with the 5 cm long single spikelet with rounded glumes apparently a continuation of the stem. (Fig. 189)

E. variegata (Poir.)Presl

Summerhayes, Bailey

Rob 2798(S)

Localities recorded: M,P

Indigenous rhizomatous sedge similar to *E. variegata* but with narrower stems, narrower spikelet and pointed glumes giving the spikelet a less smooth outline.

Fimbristylis Vahl

F. complanata (Retz.)Link (incl. var. *kraussiana* (Hochs.)C.B.Cl.) (*F. consanguinea*)

Summerhayes, Bailey

Jef 649,689(S),564(K), Fos 52054,52059(S), Pro 4110(S), Bru/War 12(S), Rob 2669,2770,2795,3105(S)

Localities recorded: M,S,P,F,N,A,Co,Coe

Tufted perennial sedge to 75 cm tall, sometimes in water, with small heads or compound inflorescences of red brown spikelets to 1 cm long on flattened stalks which extend as a bract above the inflorescence.

F. cymosa R.Br. (incl. *F. spathacea* Roth.) (*F. obtusifolia*)

Summerhayes, Bailey, Gwynne, Renvoize, Fosberg (2), Stoddart (1),(2),(3)

Ves 5385B,5930,6066(K), Jef 525(S), Pig sn(K), Sch 11749(K), Fos 52186,52191(S),52203(K), Pro 3943,4128,4129,4151(S),4391(K), Rob 2418,2516,2516A,2549,2386,2608,2643,2725,2761,2777,2806,2814,2881,3019, 3063(S)

Localities recorded: M,S,P,F,N,C,L,Co,R,Den,Pla,Coe,Rem,Des,Dar,Jos, Poi,Mar,Alp,Pie,Far,Ald,Ass,Cos, Ast

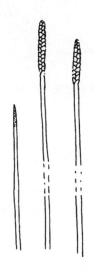

Fig. 189. Eleocharis dulcis

Fig. 190. Fimbristylis cymosa

Fig. 191. Fimbristylis dichotoma

Fig. 192. Kyllinga polyphylla

Very variable tufted perennial sedge with a rosette of leaves, often reddish green, and stiff flower stalks to 50 cm tall, with a dense head or spreading branched inflorescence of brown ovoid spikelets to 5 mm long. (*F. cymosa* has 2-and/or 3-fid stigmas, *F. spathacea* has only 2-fid stigmas) (Fig. 190)

F. dichotoma (L.)Vahl (incl. var. *laxa*) (*F. diphylla, F. annua*)

Summerhayes, Bailey, Gwynne, Renvoize

Jef 380,683,684,777,1194(S), Fos 51973(S), Pro 4211,4276(S), 4385(K), Rob 2419,2695,2815(S)

Localities recorded: M,S,P,F,L,?Coe

Tufted perennial sedge with greyish, often pubescent, leaves with a compound inflorescence of widely spaced brown spikelets, each to 1 cm long and ovoid. (Fig. 191)

F. ferruginea Vahl

Renvoize, Fosberg (2)

Localities recorded: Ald

Densely tufted sedge with erect flower stalks to 70 cm tall with up to 7 ovate lanceolate spikelets, each to 2.7 cm long.

F. littoralis Gaudich. (*F. miliacea*)

Summerhayes

Pro 4463,4494(K), Rob 2800,2818(S)

Localities recorded: M,P

Tufted annual sedge to 1 m with short leaves and an open compound inflorescence of small globose spikelets each to 2 mm long, found in wet places.

Kyllinga Rottb.
K. alba Nees

Gwynne, Renvoize (not *K. cartilaginea*), Robertson (7)

Gwy/Woo 1134,1156(K), Sch 11769(K), Pro 4141(S), Rob 2349, 3099(S)

Localities recorded: M,P,Coe,Dar,Poi

Tufted perennial sedge to 50 cm tall with solitary white globose heads to 1cm wide of small winged spikelets.

K. colorata (L.)Druce (*K. brevifolia, K. erecta*)

Summerhayes, Renvoize, Gwynne

Jef 391(S), Gwy/Woo 1107A(K), Rob 2372,3106(S)

Localities recorded: M,S,Coe,Poi

Slender creeping perennial sedge in wet places with erect stems with solitary small greenish globose heads to 5 mm wide with long drooping bracts.

K. monocephala Rottb. (*K. nemoralis, Cyperus kyllinga*)

Summerhayes, Bailey, Stoddart (1), Wilson ?(1)

Jef 536(S), Fos 52061(S), Rob 2262,2699,2625,2886,2999,3107(S)

Localities recorded: M,?P,F,N,Co,Pla,Coe,Alp,?Far

Small tufted annual or perennial sedge to 50 cm tall with flower stalks scarcely longer than the leaves and small globose solitary white heads to 6 mm wide.

K. polyphylla Willd. ex Kunth (*Cyperus aromaticus*)

Summerhayes, Bailey, Gwynne, Renvoize (not *K. melanosperma*), Fosberg (2), Stoddart (1), Wilson (1)

Jef 378(S), Gwy/Woo 892(K), Fos 51974,52058(S), Fra 443(S), Pro 4108(S),4384 (K), Bru/War 18(S), Rob 2304,2446,2557,2683,2890,3100(S)

Localities recorded: M,S,P,F,N,A,L,Co,Coe,Rem,Alp,Ald

Indigenous rhizomatous perennial sedge with red leaf sheaths and leaves half the length of the flower stalks, to 50 cm tall, with large green solitary, often 3-lobed, heads with long bracts, and spikelets with long golden stamens. (Fig. 192)

Lophoschoenus Stapf

L. hornei (C.B.Cl.)Stapf (*Schoenus xipholepis*)   'l'herbe rasoir'

Summerhayes, Bailey, Procter (1)

Ves 5436,6117(K), Pro 4468(K), Rob 2521(S)

Localities recorded: M,S,P,C

Endemic large rhizomatous perennial sedge with leaves to 1 m long, in tussocks, and erect flower stalks to 2 m long with drooping golden brown compound inflorescences of many narrowly ovate smooth spikelets to 6 mm long.

Mariscus Vahl

M. dubius (Rottb.)Fischer (*M. dregeanus, Cyperus dubius*)

Summerhayes, Bailey, Gwynne, Renvoize, Fosberg (2), Stoddart (1),(2),(3), Wilson (4), Robertson (6)

Sch 11768(K), Fos 52062(S), Pro 3917,4038,4111,4161(S) 4294(K), Fra 444(S), Bru/War 9(S), Rob 2301,2370,2565,2445,2512,2661,3047,2887,3079(S)

Localities recorded: M,S,P,F,N,A,An,L,Co,Bir,Den,Pla,Coe,Rem,Des,Dar, Poi,Mar,Alp,Far,Ald,Ass

Indigenous slender tufted sedge to 50 cm tall with pseudobulbs and solitary globose heads to 1 cm wide of white spikelets, each to 5 mm long. (Fig. 193)

269

M. ligularis (L.)Urb. (*M. rufus, Cyperus ligularis*)

Summerhayes, Bailey, Renvoize, Fosberg (2), Stoddart (1),(2),(3), Wilson (3),(4)

Squ G142(K), Ves 5402,6150(K), Jef 1201(S), Fos 52036(S),52119(K), Pro 4107,4175,4225(S), Fra 442,456(S), Bru/War 11(S), Rob 2340,2513, 2878,2892,3041,3055(S)

Localities recorded: F,A,L,Co,Cu,R,Se,V,Bir,Den,Pla,Coe,Afr,Rem,Des,Dar,Jos, Poi,Mar,Def,Alp,Pro,Far, Ald,Ass,Cos,Ast

Tall stout sedge to 1 m with long grey green leaves and stiff stems with a compound inflorescence of densely arranged rusty brown oblong spikes to 2 cm long of ovate spikelets each to 5 mm long.

M. paniceus (Rottb.)Vahl

Summerhayes, Bailey

Jef 632(K), Pro 4138(S),4471,4507(K), Rob 2477(S)

Localities recorded: M,S,P,D,Fe

Indigenous tufted sedge to 50 cm tall with erect stems with a compound inflorescence of a few oblong spikes to 2 cm long with slight shiny golden green spikelets each to 4mm long.

M. pedunculatus (R.Br.)Koyama (*Remirea maritima*)

Dup 202(K), Lam 420(K), Pro 3942,4140(S), Rob 2727(S), 2771(K), Tod n/n(S)

Localities recorded: M,P,F,A

Sedge of beach crest with deep rhizomes and seasonal rosettes of stiff leaves with short stems and dense lobed heads of small white spikelets, each to 5 mm long, which dry stiff and brown.

M. pennatus (Lam.)Domin. (*Cyperus javanicus*)

Summerhayes

Sch 11747(K), Fos 52037(S), Pro 4289(S),4386(K)

Localities recorded: M,S,P,C,Cu

Indigenous tall tufted sedge to 1 m near swamps, with erect stems with large compound inflorescence of loose cylindrical spikes to 4 cm long of pale brown ovoid spikelets, each to 8 mm long and separate from each other.

M. tenuifolius Schrad.

Pro 4477(K)

Localities recorded: M(BG)

Slender tufted sedge to 30 cm tall with narrow leaves and spikes in a V-shaped cluster to 1.5 cm long, the spikelets narrow and pale green, each to 6 mm long. Possibly introduced from Sri Lanka.

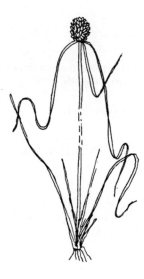

Fig. 193. Mariscus dubius

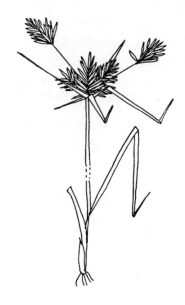

Fig. 194. Pycreus polystachyos

Fig. 195. Schoenoplectus
juncoides

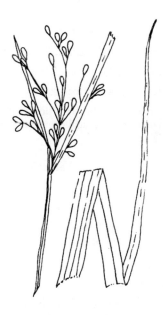

Fig. 196. Thoracostachyum
floribundum

Pycreus Beauv.

P. globosus (All.)Reichb. (*Cyperus globosus, C. flavidus*)

Summerhayes, Bailey

Jef 392(S), Rob 2794(S)

Localities recorded: M,P

Slender tufted sedge to 50 cm in damp places with compound inflorescences of a few long slender flattened spikelets, each to 1.5 cm long and silvery green.

P. polystachyos (Rottb.)P.Beauv. (*P. odoratus, Cyperus polystachyos*)

Summerhayes, Bailey, Gwynne, Renvoize

Sch 11748(K), Pig sn(K), Fos 52038,52087(S), Pro 3993,4113(S), Bru/War 10(S), Rob 2263,2476A,2611,2799,3101(S)

Localities recorded: M,S,P,F,N,A,Co,Den,Coe,Pie

Tufted sedge to 50 cm tall with stalks much longer than the leaves and compound heads of an inverted pyramidal shape of many long narrow green or golden brown spikelets each to 1.5 cm long. (Fig. 194)

P. pumilus (L.)Vahl (*Cyperus pumilus*)

Fosberg (2)

Localities recorded: Ald

Slender tufted annual sedge to 25 cm tall with loosely globose or compound heads of linear flat spikelets, each to 1.5 cm long.

Schoenoplectus Palla

S. juncoides (Roxb.)Palla (*Scirpus juncoides*)

Rob 2817(S)

Localities recorded: M

Apparently leafless perennial sedge in standing water with tall rounded stems with a few fat spikelets, each to 1 cm long, in a cluster up to 10 cm below the tip. (Fig. 195)

S. mucronatus (L.)Palla (*Scirpus mucronatus*)

Summerhayes

No locality recorded

Similar to *S. juncoides* but with wider stems, more spikelets per cluster, and clusters nearer the tip of the stem. This is possibly a misidentification for *S. juncoides*.

Scleria Berg.

S. sieberi Nees ex Kunth (*S. angusta* var. *seychellensis*)   'herbe coupant'

Summerhayes, Bailey, Evans

272

Jef 541,648,727(S), Ber 14650(K), Rob 2479(S)

Localities recorded: M,S,P

Indigenous perennial sedge with clumps to 1.5 m tall of slender leafy stems, triangular in section, with narrow compound inflorescences in the leaf axils, of many untidy red brown spikelets containing black or white grains in brown 'cups'.

S. sumatrensis Retz. 'herbe coupant'

Summerhayes, Bailey

Jef 1229(S), Sch11760(K), Pro 4267(S), Rob/Wil 2479(S)

Localities recorded: P,C

Indigenous perennial sedge with clumps to 3 m tall of stems triangular in cross section with short stiff leaves and untidy inflorescences of pale greenish brown spikelets and grey or white grains held in a red 'cup'.

Thoracostachyum Kurz

T. angustifolium C.B.Cl.

Summerhayes, Procter (1),(2)

Jef 1228(S), Pro 4029,4217(S)

Localities recorded: M,S,P

Endemic tufted perennial sedge to 1 m tall with narrow scabrid leaves and compound inflorescences of a few obovate spikelets each to 3 mm long.

T. floribundum (Nees)C.B.Cl. (?Hypolytrum seychellensis)

Summerhayes, Procter (1),(2)

Ves 5418,5444,5528(K), Jef 416(K), Pro 3977,4237(S), Rob 2470,2633(S)

Localities recorded: M,S,P,D,C,Fe

Endemic tufted perennial sedge to 2 m tall with wide leaves and compound inflorescences of many widely spaced obovate spikelets each to 6 mm long. (Fig. 196) (These species may now be in *Costularia*).

GRAMINEAE (200)

Aristida L.

A. depressa Retz. (A. mauritiana)

Summerhayes

No locality recorded

Slender grass to 25 cm tall, similar to A. *setacea* but with awns to 2 cm long and shorter leaves.

A. setacea Retz.

Dup 2/31(K), Pro 4502(K), Rob 2432(S)

Localities recorded: M

Slender grass to 1 m in wet places with tall feathery inflorescences with long thin spikelets, the awns to 4 cm long, and the leaves often reflexed.

Arundo L.

A. donax L.

Robertson (7)

Rob 3162,3163(S)

Localities recorded: M,Dar

Ornamental giant perennial grass in clumps to 5 m tall with glaucous green, or yellow and white variegated, leaves, and large feathery white inflorescences with long white hairs on the spikelets. The stems make good canes for plant supports.

Axonopus P.Beauv.

A. compressus (L.)P.Beauv.

Summerhayes

Jef 808(S), Sch 11785(K), Pro 4525(K), Rob 2399,2598,2665(S)

Localities recorded: M,S,F,N

Creeping grass with pinkish stolons, flattened stems and folded leaves and inflorescences of 2 or 3 digitately arranged spikes up to 8 cm long with small green spikelets in 2 rows. A lawn and pasture grass.

Bambusa Schreb.

B. arundinacea (Retz.)Willd.   'bambou, bamboo'

Bailey

No locality recorded

Clumped bamboo to 30 m tall with branching green stems to 15 cm wide and leaf blades to 20 cm long and 2.5 cm wide.

B. glaucescens (Willd.)Sieb. ex Munro (*Bambusa nana*)   'bambou nain, herbe bambou'

Bailey

No locality recorded

Clumped bamboo to 5 m with leaf blades to 12 cm long and 1.5 cm wide.

B. vulgaris Schrad. ex Wendl. (incl. var. *aureo-variegata*)   'bambou jaune'

Bailey, Fosberg (1),(2), Wilson (1)

Bru/War 54(S), Tod 11(S)

Localities recorded: P,F,A,Ald

Erect bamboo to 5 m or more, stems and young branches often black, a distinct ridge beneath each node, leaf blades lanceolate, pointed, to 30 cm long, and a large leafy inflorescence.

Bothriochloa Kuntze

B. pertusa (L.)A.Camus

Pro 4559(K)

Localities recorded: M

Tufted perennial grass to 50 cm tall, rooting at the nodes, leaves mostly basal, with 2 to 7 racemes, almost digitately arranged, softly hairy to 7 cm long, and spikelets to 4 mm long with white hairs to 2 mm long at the base, the ♀ spikelet with a pit at the back.

Brachiaria Griseb.

B. brizantha (A.Rich.)Stapf

Rob 2525(S)

Localities recorded: M

Erect tufted grass to 50 cm tall with several widely spaced racemes up to 8 cm long held at right angles to the stems and each raceme with two rows of fat ovoid spikelets, each to 5 mm long.

B. mutica (Forssk.)Stapf   'para grass, mauritius grass'

Summerhayes

Localities recorded: M

Straggling grass to 3 m long, rooting at the nodes, with many spreading racemes on each inflorescence, the spikelets clustered or solitary, ovate, to 3.5 mm long.

B. subquadripara (Trin.)Hitchc.

Fos 52072(S), Rob 2525,2719,2885,3154(S)

Localities recorded; F,Co,Coe,Alp

Straggling grass, rooting at the nodes, with slender stems with a few widely spaced racemes up to 5 cm long, each with one row of green ovate spikelets, each to 4 mm long.

B. umbellata (Trin.)W.D.Clayton (*Panicum nossibense, Panicum parvifolium, Panicum umbellatum*) 'gazon trelle, gazon chinensis, herbe edwards'

Summerhayes, Vesey Fitzgerald (1), Bailey, Swabey, Wilson (1), Robertson (6), Sauer

275

Jef 1193(S),382(K), Sch 11820(K), Fos 51964(S), Rob 2486,2982(S)

Localities recorded: M,P,F,C,An,Dar

Indigenous mat-forming grass, often used for lawns, rooting at the nodes and sometimes forming a deep cushiony sward, with stems to 15 cm tall and small panicles of an inverted pyramid-shape, with small green ovate spikelets to 1.5 mm long. (Fig. 197)

Cenchrus L.

C. echinatus L.

Renvoize, Fosberg (2), Stoddart (1),(2),(3), Robertson (3)

Jef 1202(S), Gwy/Woo 172(K) (not *C. mitis* Anders.), Fos 52160(S), Fra 428(S), Rob 2309,2977,3128(S)

Localities recorded: Co,Den,Coe,Rem,Dar,Poi,Far,Ast

Erect tufted grass to 50 cm tall with hairy leaf blades and a spike of burred spikelets, each spikelet surrounded by whorls of barbed bristles. (Fig. 198)

Chloris Sw.

C. barbata (L.)Sw.

Summerhayes, Renvoize, Stoddart (1) Squ G140(K), Jef 388(S), Pro 4092(S), Rob 2416, 2562,2739,2989(S)

Localities recorded: M,F,N,L.Dar,Far

Slender grass to 50 cm tall with digitately arranged racemes to 8 cm long with 3-awned spikelets arranged on one side, sometimes purplish.

C. pycnothrix Trin.

Jef 526(S), Pro 3966(S)

Localities recorded: M

Tufted grass similar to *C. barbata* but with wider and longer leaf blades and spikelets with 2 awns, one to 1.8 cm long, on curved racemes.

C. virgata Sw.

Squ G11(K)

Localities recorded: L

Tufted grass similar to *C. pycnothrix* and spikelets with 2 awns, but the longer to 1 cm long only and with long white hairs between them.

Chrysopogon Trin.

C. aciculatus (Retz.)Trin.

Summerhayes

Jef 703(S), Pro 4215(S), Rob 2489(S)

Fig. 197. Brachiaria umbellatum

Fig. 198. Cenchrus echinatus

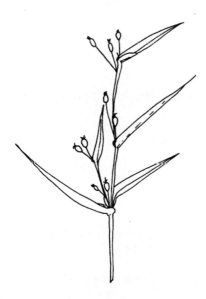

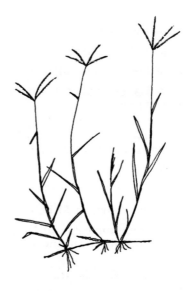

Fig. 199. Coix lachryma-jobi

Fig. 200. Cynodon dactylon

Localities recorded: M

Creeping grass with short leaves and erect stems to 20 cm tall bearing a feathery panicle of slender mauve tinted spikelets.

Coix L.

C. lachryma-jobi L.   'herbe collier, herbe job, job's tears'

Summerhayes, Bailey, Baumer

Tod 54(S)

Localities recorded :F

Cultivated and escaping robust annual to 2 m with wide leaves and the spikelet enclosed in a hollow grey bead-like structure. Used to make beads, and used medicinally in Seychelles. (Fig. 199)

Cymbopogon Spreng.

C. citratus (DC.)Stapf   'citronelle petit, lemon grass'

Bailey, Lionnet (1), Fosberg (2), Evans

Dup 9/34(K), Rob 2538(S)

Localities recorded: M,P,Ald

Tufted perennial grass to 2 m with a long stalked large compound inflorescence to 60 cm long of unawned spikelets with yellow hairs in short racemes which are subtended by brown spathes to 1.5 cm long, the lower glume of the spikelet slightly depressed below the middle on the back. The leaves yield west indian lemon grass oil, and are used for tea.

C. flexuosus (Nees)W.Wats.   'lemon grass(poor quality)'

Dup 6/34,5/34(K)

Localities recorded: M(BG)

Tufted grass similar to *C. citratus* but with lower spikelet with a short bent awn to 5 mm long.

C. martinii (Roxb.)Wats. (incl. var. *motia* and var. *sofia*) 'rosha grass, gingergrass (var. *sofia*), palmarosa (var. *motia*)'

Bailey

Dup 2?/33,3/33,14/34,13/34(K)

Localities recorded: M(BG)

Tufted perennial grass to 2 m or more, similar to *C. pruinosus* but with longer leaf blades, to 50 cm, and shorter florets, the glume winged.

C. nardus (L.)Rendle (incl. var. **confertiflorus**)   'citronelle grosse, citronella grass'

Bailey

Dup 7/34,8/34,10/34,11/34,12/34(K)

Localities recorded: M(BG)

Robust stoloniferous perennial grass to 1 m tall with broad leaves, a large compound panicle with awnless spikelets similar to *C. citratus* but with lower glume flat on the back. The leaves yield citronella oil and Dupont called his 8/34 Ceylon citronella, and his 10/34 Java citronella.

C. pruinosus (Nees ex Steud.)Chiov.

Pro 4357(K)

Localities recorded: S

Tufted grass to 50 cm with leaf blades to 30 cm long and rounded at the base, a compound panicle to 25 cm long and pairs of racemes on short stalks with a large bract, to 2.5 cm long, lower spikelet with a bent awn to 1.8 cm long, pale above and dark below, and the lower glume 3 toothed.

Cynodon Rich.

C. dactylon (L.)Pers.   'chiendent, bermuda grass, ?patte de poule'

Summerhayes, Bailey, Stoddart (1),(2),(3)

Jef 753(S), Pro 4250,4157(S), Bru/War 16(S), Rob 2511,2721,3002,3089(S)

Localities recorded: M,F,A,Bir,Pla,Coe,Des,Dar

Creeping grass with wiry stems and short glaucous leaves, erect stems to 20 cm long, usually less, with 4 to 5 digitately arranged racemes up to 6 cm long with spikelets up to 2 mm long arranged on one side. (Fig. 200)

Cyrtococcum Stapf

C. oxyphyllum Stapf

Summerhayes

Jef 795(S), Pro 4504,4561(K), Rob 2447(S)

Localities recorded: M,S,P

Indigenous straggling grass in woodland glades with small round spikelets in a loose panicle with all the branches held erect.

C. trigonum (Retz.)A.Camus

Pro 4153(S), Rob 2693(S)

Localities recorded: F

Prostrate mat-forming grass similar to *Brachiaria umbellatum*, but with pubescent, one-sided spikelets.

Dactyloctenium Willd.

D. ctenoides (Steud.)Bosser (*D. aegyptium*)   'herbe bourrique, chiendent patte de poule'

Summerhayes, Bailey, Gwynne, Renvoize, Fosberg (2), Stoddart (1),(2),(3), Wilson (3),(4), Robertson (6)

Pig sn, sn(K), Sch 11757(K), Fos 52177(S), Pro 3944,3908,4227(S), 4403(K), Fra 424(S), Bru/War 86(S), Rob 2244,2331,2368,2606,2722,2843,2847,2897, 2978,3037,3058,3139(S)

Localities recorded: M,S,P,F,N,A,An,L,Co,Bir,Den,Pla,Coe,Afr,Rem,Des, Dar,Jos,Poi,Mar,Def,Alp,Pie,Far, Ald,Ass,Cos,Ast

Erect or spreading annual or perennial grass, rooting at the nodes and with stems to 20 cm long bearing several short stout spikes, digitately arranged and usually held at right angles to the stem, with tightly packed spikelets in one row on each spike. Said to differ from *D. aegyptium* (L.) Willd. in the grain being less transversely rugose. (Fig. 201)

D. pilosum Stapf

Summerhayes, Renvoize, Fosberg (2)

Ves 5943A(K)

Localities recorded: P,Coe,Pie,Ald,Ass,Cos,Ast

Indigenous spreading annual or perennial grass, rooting at the nodes with stems to 25 cm long bearing 2 short (to 3.7 cm long), somewhat curved, spikes with crowded spikelets, each to 3 mm long, on one side only.

Daknopholis W.D.Clayton

D. boivinii (A.Camus)W.D.Clayton

Renvoize, Fosberg (2)

Localities recorded: Ald,Cos,Ast

Spreading annual grass at beach crest with inflorescences of up to 4 slender racemes, each to 6 cm long, with spikelets, each with an awn to 1 cm long, along one side only.

Dendrocalamus Nees

D. giganteus Munro   'bambou geant'

Bailey, Lionnet (2)

?Rob 2749A(S), ?Tod 18(S)

Localities recorded: M(BG),F

Giant bamboo with stems to 15 cm wide.

D. strictus (Roxb.)Nees   'bambou gaulette, bambou plein, male bamboo'

Bailey

No locality recorded

Giant bamboo with stems to 7.5 cm wide.

Dichanthium Willem.

D. aristatum (Poir.)Hubb.

Pro 3995(S)

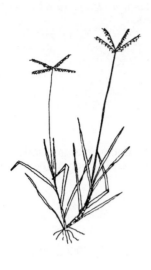

Fig. 201. Dactyloctenium ctenoides

Fig. 202. Digitaria horizontalis

Fig. 203. Echinochloa colonum

Fig. 204. Eleusine indica

281

Localities recorded: S

Slender perennial grass with stems to 75 cm with solitary or paired spikes of hairy awned spikelets to 8 cm long. Similar to *Bothriochloa pertusa* but without pit at back of spikelets and with very short hairs.

Digitaria Fabr.

D. ciliaris (Retz.)Koel.

Stoddart (2)

Pro 4277(S)

Localities recorded: P,Dar

Spreading grass with inflorescence of up to 7 digitately arranged spikes, with spikelets to 3.5 mm long with minute lower glume and upper glume nearly as long as spikelet.

D. didactyla Willd. (*D. pruriens*)

Summerhayes (in part)

Gardiner n/n A &C(K), n/n(K), Thomasset n/n(K), Dup 206(K), Pro 4055(S), Rob 3097(S)

Localities recorded: M,Fe,L,Coe

Indigenous mat-forming grass with slender erect stems to 30 cm tall bearing 2 racemes, rarely a short third, of small green untidy spikelets, each to 3 mm long, in one row, with minute lower glume and upper glume more than half the length of the spikelet.

D. horizontalis Willd.

Summerhayes, Gwynne, Fosberg (2), Stoddart (1),(2),(3)

Squ G7,G174(K), Fos 52066(S), Pro 4049,4118,4125(S), 4170(K), Bru/War 7,27(S), Rob 2371,2509,2685,2896(S)

Localities recorded: M,S,P,F,A,L,Co,Bir,Des,Dar,Poi,Ald,Cos,Ast

Vigorous straggling grass with slender erect stems to 60 cm tall with long slender racemes to 15 cm long, arranged in whorls on a short axis, with 2 or 3 rows of slender spikelets with a minute lower glume and upper glume half the length of the spikelet. (Fig. 202)

D. radicosa (Presl)Miq. (*D. timorensis*)

Summerhayes (in part from *D. pruriens*), Gwynne, Stoddart (1)

Gardiner n/n,B&D(K), Fos 52040,52175(S), Pro 3916(S), Rob 2570(S), Tod n/n(S)

Localities recorded: M,S,F,N,A,Co,Cu,Coe,Far

Slender straggling grass, similar to *D. horizontalis*, rooting at the nodes, but with fewer racemes, the spikelets to 3 mm long and the lower glume minute or missing, the upper glume up to half the length of the spikelet.

D. scalarum (Schweinf.)Chiov.

Bailey

No locality recorded

Probably a misidentification by Bailey who interpreted 'herbe calume' as 'couch grass'. All 'herbe calume' I saw was *Panicum repens*.

D. setigera Roth (*D. aegyptiaca*, *D. marginata*)   'herbe gros gazon, herbe perard grande'

Summerhayes, Bailey, Fosberg (2), Wilson (4)

Gwy/Woo 1124,1139,1202(K), Sto/Poo 1357,1414(K), Fos 52101(S), Pro 4212,3965(S),4330,4509(K), Rob 2245,3013,3066(S)

Localities recorded: M,P,Co,Pla,Coe,Des,Mar,Far,Ald,Ast

Straggling grass similar to *D. horizontalis* with spikelets with no lower glume and upper glume less than a quarter the length of the spikelet.

Note — *D. ciliaris*, *D. horizontalis*, *D. radicosa* and *D. setigera* are similar and the localities quoted may be incorrect.

Echinochloa P. Beauv.

E. colonum (L.)Link

Summerhayes

Jef 657(S), Pro 4358,4366(S), Bru/War 38(S), Rob 2555, 2629(S)

Localities recorded: M,S,N,A

Tufted annual grass to 60 cm in wet places with inflorescence of up to 10 short racemes held at an angle to the axis and crowded with several rows of short fat apiculate spikelets, often purplish green. (Fig. 203)

Eleusine Gaertn.

E. indica (L.)Gaertn.   'chiendent patte de poule, herbe patte de poule'

Summerhayes, Bailey, Gwynne, Renvoize, Fosberg (2), Stoddart (1),(2),(3), Baumer, Wilson (1),(3),(4), Robertson (6)

Lionnet 2(K), Jef 527,712(S), Fos 52098(S), Pro 4048,4119,4126(S),4324(K), Rob 2332,2360,2265,2602,2696,2860,3016,3122(S),2508(K)

Localities recorded: M,S,P,F,N,A,An,Co,Bir,Den,Pla,Coe,Des,Dar,Mar,Alp, Far,Ald,Cos,Def,Poi

Tough erect annual tufted grass to 85 cm tall with an inflorescence of up to 10 digitately or subdigitately arranged spikes, each to 15 cm long with several rows of untidy spikelets. Used medicinally in Seychelles. (Fig. 204)

Enteropogon Nees

E. sechellensis (Baker) Dur. & Schinz (*E. melicoides*, all Seychelles material referred to as *E. monostachyos*, *Ctenium sechellense*)   'gazon'

Summerhayes, Bailey, Renvoize, Fosberg (2), Gwynne, Stoddart (1),(3), Robertson(6)

Jef 1200(S), Pig sn(K), Fos 52172(S), Pro 3952,4109 (S),4499(K), Bru/War 15(S), Rob 2466,2893(S),2500(K)

Localities recorded: P,F,A,An,Co,Bir,Den,Des,Dar,Pie,Ald,Ass,Cos,Ast

Indigenous erect tufted perennial with wiry stems to 1 m with one curved spike to 20 cm with a double row of large awned spikelets on one side.

Eragrostis Wolf

E. ciliaris (L.)R.Br.

Summerhayes, Gwynne, Renvoize, Stoddart (2)

Pro 4169(S),4497(K), Rob 2973,3071(S)

Localities recorded: P,F,Coe,Dar,Jos

Tufted annual grass to 40 cm, with short leaves and a subspike of densely arranged woolly purplish spikelets, each to 15 cm long.

E. decumbens Renvoize

Renvoize, Fosberg (2)

Localities recorded: Ald,Ass,Cos,Ast

Endemic wiry perennial grass with a rosette of narrow leaves to 4 cm long and a subspike with broad spikelets to 4 cm long.

E. pilosa (L.)P. Beauv.

Summerhayes

No locality recorded

Slender annual to 60 cm with a delicate panicle with slender spikelets on long stalks and long tufts of hair in the lowest axils.

E. racemosa (Thunb.)Steud.

Rob 2752(S)

Localities recorded: M

Slender annual grass with short leaves and stems in a rosette, spreading at first, then erect, with an open panicle with greyish green spikelets, each to 5 mm long, with persistent glumes.

E. subaequiglumis Renvoize

Summerhayes (*E. riparia* in part), Renvoize, Fosberg (2), Stoddart (3), Wilson (4)

Gardiner n/n(K), Ves 5656(K), Gwy/Woo 981(K), Sch 11777(K), Fos 52176,52095(S), Pro 4044(S), Fra 433(S), Rob 2325,2379,2882,2898, 2974,3018,3072,3073(S)

Localities recorded: P,Co,Bir,Pla,Coe,Rem,Des,Jos,Dar,Poi,Mar,Alp,Far, Ald,Ass,Cos,Ast

Tufted annual or perennial grass to 40 cm with long narrow inrolled leaves and wiry stems, often in a rosette, with a contracted panicle of appressed branchlets of almost sessile long narrow pale green, drying cream, spikelets, each to 6 mm long. (Fig. 205)

E. tenella (L.)P.Beauv.

Summerhayes, Gwynne, Renvoize, Stoddart (1),(3)

Jef 1199(S), Fos 52167(K), Pro 4123(S),4503(K), Bru/War 17(S), Rob 2347,2260,2975,3078(S)

Localities recorded: M,S,D,A,L,Co,Bir,Den,Coe,Des,Dar,Poi

Slender tufted grass to 50 cm with short flat leaf blades and delicate open panicle with widely spaced small green, drying cream, spikelets, each to 3 mm long. (Fig. 206)

E. tenella (L.)P.Beauv. var. insularis Hubb.

Summerhayes (*E.riparia* in part), Gwynne, Stoddart (1),(2)

Gardiner 24,n/n,n/n(K), Thomasset n/n(K), Horne 232,n/n(K), Dup 22,40(K), Squ G8(K), Jef 686,754(S), Pro 2967,4124,4165(S),4506(K), Rob 2600,2743(S)

Localities recorded: M,S,P,F,N,L,Den,?Dar,?Jos

Slender tufted grass to 50 cm with short flat leaf blades and contracted spike-like panicle with small green spikelets, each to 3 mm long, on appressed branchlets. Difficult to distinguish from *E. subaequiglumis* and the localities for each may be incorrect.

Eriochloa Kunth

E. meyeriana (Nees)Pilger

Renvoize, Fosberg (2)

Localities recorded: Ald,Ass

Loosely tufted rhizomatous perennial grass to 1 m, with a panicle to 16 cm long with purplish ovate spikelets to 3.5 mm long, thickly and untidily arranged on the branchlets.

E. subulifera Stapf

Renvoize, Fosberg (2)

Localities recorded: Ass,Ast

Delicate annual grass to 70 cm tall with a panicle to 12 cm long with ovate spikelets to 2 mm long with the glume with a distinct point, thickly arranged on the branchlets.

Garnotia Brongn.

G. sechellensis Hubb. & Summerhayes

Summerhayes, Procter (1)

Jef 1142(S), Pro 4476,4488(K), Rob 2701(S)

Localities recorded: M,S

Endemic rare tufted or straggling grass to 1.5 m tall with long wide leaves and a loose panicle to 25 cm long, of narrow pale green spikelets, each to 6 mm long.

Heteropogon Pers.

H. contortus (L.)P.Beauv.

Summerhayes, Robertson (6)

Dup 1/31(K), Squ G21,G89(K), Jef 690(S), Rob 2439(S)

Localities recorded: M,An,L

Tufted perennial grass to 1 m with a solitary raceme to 15 cm long with the upper spikelets with awns to 10 cm long which twist together. (Fig. 207)

Hyparrhenia Anders.

H. rufa (Nees)Stapf

Summerhayes

Ves 5438,5630,5613(K), Jef 366(K), Pro 3947(S), Rob 2461 (S)

Localities recorded: M,S,P,C

Clumped perennial or tufted grass to 2.5 m tall with a loose panicle of pairs of short racemes on a long stalk to 6 mm or more, subtended by a reddish bract, the lower spikelet with the awn bent halfway, black below and straw-coloured above.

Ischaemum L.

I. heterotrichum Hack.   'herbe pilaute'

Summerhayes

Gardiner 36(part)(K), Ves 6437(K), Jef 1239(K), Sch 11750(K), Pro 4050,4234(S),4498(K), Bru/War 13(S), Rob 2398(S)

Localities recorded: M,P,A,L

Creeping mat-forming grass, often near the sea, with pointed leaves to 15 cm long and paired racemes of large smooth awned spikelets, each to 8 mm long with the awn to 1.4 cm long, the pairs held together and often infected with black smut.

I. indicum (Houtt.)Merr.

Sch 11730(K)

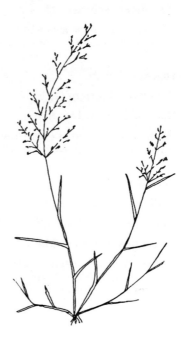

Fig. 205. Eragrostis subaequiglumis   Fig. 206. Eragrostis tenella

Fig. 207. Heteropogon contortus   Fig. 208. Lepturus repens

Localities recorded: M

Similar to *I. heterotrichum* but with spikelets to 4 mm long, the awn to 1 cm long and more densely arranged on the spike.

I. rugosum Salisb.

Renvoize, Fosberg (2)

Localities recorded: Ald

Robust annual grass with narrow leaves and paired racemes with awned spikelets with transverse wrinkles on the lower glume.

Lepturus R.Br.

L. radicans (Steud.)Camus   'chiendent'

Dup 13/33(K), Ves 6180(K), Jef 1182(S), Pro 4475(K), Rob 2582(S)

Localities recorded: M,P,F,N

Creeping grass with the inflorescence a solitary spike with alternate spikelets sunk in the hollowed side of a swollen axis, giving a zigzag effect, the spikelets to 3 mm long, breaking off when mature.

L. repens (G.Forst.)R.Br.

Summerhayes, Gwynne, Renvoize, Fosberg (2), Stoddart (1),(2),(3), Wilson (3),(4), Robertson (3)

Ves 5808(K), Fos 52042(S), Pro 4229(S), Rob 2491,2872,3025,3054(S)

Localities recorded: Cu,Bir,Pla,Coe,Afr,Rem,Dar,Jos,Poi,Mar,Def,Alp,Far, Ald,Cos,Ast

Indigenous stoloniferous perennial of the beach crest with the inflorescence a solitary spike with spikelets sunk in the swollen axis alternately, the upper glume to 1.5 cm long and projecting at an angle from the rachis. The spike breaking up into segments when mature. (Fig. 208)

Oplismenus P.Beauv.

O. compositus (L.)P.Beauv.

Summerhayes, Robertson (6)

Squ G13(K), Jef 790(S), Pro 3915,3919(S), Rob 2414,2593(S)

Localities recorded: M,S,P,N,An,L

Indigenous straggling grass, usually in woodland, with crinkled leaf blades and the inflorescence of several short distant racemes on which the spikelets, each to 4mm long and with an awn to 1.5 cm long, are also distant, the spikelets falling early.

Oryza L.

O. sativa L.   'riz, rice'

Bailey

No locality recorded

Freely tillering annual grass to 1.5 m tall, but longer in floating varieties, with terminal panicle with many creamy yellow stiff spikelets, each to 8 mm long, containing an edible white grain.

Panicum L.

P. aldabrense Renvoize

Renvoize, Fosberg (2)

Localities recorded: Ald

Endemic small stoloniferous perennial to 5 cm tall, with small inflorescences to 1 cm long, and spikelets to 1.3 mm long.

P. assumptionis Stapf

Renvoize, Fosberg (2)

Localities recorded: Ass

Endemic grass similar to *P.aldabrense* but with stouter stolons which are semi-erect, with clusters of leaves and inflorescences to 1.5 cm long.

P. brevifolium L.   'herbe la seine'

Summerhayes, Bailey, Robertson (6)

Ves 5437,5615,5629(K), Squ G2(K), Jef 353(S), Sch 11710,11786(K), Fos 52050(S), Pro 4036,4116(S), Ber 14634(K), Rob 2586,2640(S)

Localities recorded: M,P,F,N,C,A,An,Co

Indigenous delicate spreading grass, usually in deep shade in woodland, with short wide ovate leaves and wide open panicle with small unequal-sided spikelets, each to 2 mm long. (Fig. 209)

P. maximum L.   'fataque, herbe de guinee, guinea grass'

Summerhayes, Bailey, Gwynne, Fosberg (2), Stoddart (1),(2),(3), Wilson, (1),(4), Robertson (2),(3),(6)

Jef 1175(S), Fos 52105(S), Bur/War 80(S), Rob 2490,2686(S)

Localities recorded: M,S,P,F,An,L,Co,A,De,Dar,Poi,Mar,Far,Ald,Ass,Cos,Ast

Vigorous tufted perennial grass to 2.5 m tall with long narrow leaves and a large pyramidal open inflorescence to 70 cm long with plump ovate spikelets, each to 3.3 mm long. Used as fodder grass and to make brooms.

P. repens L.   'calume, herbe calume, herbe siflette'

Summerhayes, Stoddart (3)

Pro 4440,4455(K), Rob 2413,2751(S)

Localities recorded: M,V,Bir

Rhizomatous perennial grass, a serious weed, with long white underground stolons and erect stems to 1 m tall with rigid glaucous green leaf blades and a large panicle to 20 cm long with pale pinkish spikelets, each to 3 mm long.

P. voeltzkowii Mez

Renvoize, Fosberg (2)

Localities recorded: Cos,Ast

Tufted perennial grass with narrow leaf blades to 8 cm long and a finely branched spreading panicle to 10 cm long with small ovoid spikelets, each to 1.7 mm long.

Paspalidium Stapf

P. geminatum (Forsk.)Stapf

Summerhayes

Jef 718(S), Pro 4174,4375(S), Rob 2526,2610,2645(S)

Localities recorded: M,S,F,N

Indigenous perennial grass in wet places to 1.5 m tall, with a long inflorescence of widely separated, but appressed to the axis, short racemes, each to 3 cm long, with two distinct rows of plump ovate spikelets, each to 2 mm long.

Paspalum L.

P. conjugatum Berg   'buffalo grass'

Summerhayes, Wilson (1)

Jef 365(S), Pro 4210(S), Rob 2256,2671(S)

Localities recorded: M,P,F

Indigenous slender stoloniferous perennial with paired racemes held horizontally, at right angles to the axis, each to 12 cm long and bearing a neat double row of flat round spikelets, each to 2 mm long, the tip of the raceme often bare.

P. nutans Lam. (*P. mandiocanum, P. protensum*)   'herbe perard'

Summerhayes, Bailey

Jef 354(S), Pro 4028,4209(S),4519(K)

Localities recorded: M,P

Tufted grass to 50 cm tall with solitary racemes to 4 cm long with a double row of flat round spikelets, each to 2 mm long.

P. paniculatum L.   'herbe ma tante'

Bailey

Fig. 209. Panicum brevifolium

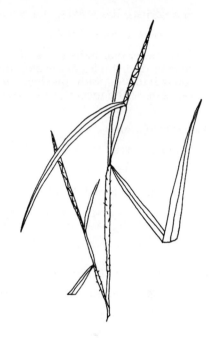

Fig. 210. Rottboelia exaltata

Fig. 211. Sporobolus virginicus

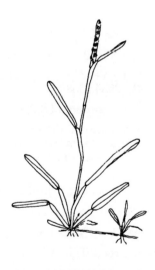

Fig. 212. Stenotaphrum dimidiatum

Squ G25(K), Ves 5582(K), Jef 364,685(S), Sch 11788(K), Rob 2417(S)

Localities recorded: M,P,L

Erect perennial hairy grass with inflorescence of many spike-like racemes held at right angles to the axis, the lower to 10 cm long and getting shorter towards the top, with spikelets, each to 2 mm long, crowded in 3 or 4 rows on one side of the raceme, the spikelets round and flattened on one side.

P. scrobiculatum L.

Summerhayes

Squ G5(K), Pro 4171,4208,4372(S), Rob 2401,2558,2664(S)

Localities recorded: M,S,N,F,L

Tall rhizomatous perennial grass with usually 2 racemes, one below the other, to 10 cm long, each with 2 neat rows of round flat spikelets, each to 2.5 mm long.

P. vaginatum Sw. (*P. distichum*)

Gwynne, Sauer, Renvoize, Fosberg (2), Stoddart (1)

Pro 3945,4251(S)

Localities recorded: P,V,Afr,Far,Ald,Cos

Rhizomatous perennial grass of the sea shore, with stems to 70 cm tall and the inflorescence of two terminal racemes, each to 6 cm long, with 2 rows of untidy oblong pointed spikelets each to 4.5 mm long.

Pennisetum Rich.

P. polystachyon (L.)Schult.   'herbe ma tante'

Summerhayes, Bailey, Gwynne, Renvoize, Fosberg (2), Stoddart (1),(2),(3)

Squ G17(K), Ves 5581(K), Jef 523(S), Sch 11729(K), Bru/War 24(S), War 193(S), Rob 2415,2554,2688(S)

Localities recorded: M,S,P,F,N,A,L,Bir,Den,Dar,Far,Ald,Ass,Ast

Erect tufted annual or perennial grass to 1.5 m tall with a solitary spike-like inflorescence 25 cm long, crowded with purplish spikelets, each to 5 mm long, and hairy bristles to 20 mm long, soft like a cat's tail.

P. purpureum Schum.   'herbe elephant, elephant grass'

Bailey, Sauer, Stoddart (1), Robertson (4)

Jef 383(S), Gwy/Woo 1107B(K), Bru/War 120(S)

Localities recorded: M,F,A,Coe,Far

Robust perennial grass to 6 m tall with long flat leaf blades and a dense spike-like inflorescence to 30 cm long with pale golden spikelets, each to 7 mm long and bristles to 4 cm long, like a long thin bottle-brush.

Perotis Ait.

P. hildebrandtii Mez (*P. indica*)

Summerhayes

Squ G92(K)

Localities recorded: M,L

Loosely tufted annual grass with short leaves and slender cylindrical spike to 40 cm tall with delicate narrow purple awned spikelets, each to 4 mm long with awns to 1 cm long, held at right angles to the spike.

Phyllostachys Sieb. & Zucc.

P. nigra (Lodd.)Munro (*Bambusa nigra*)   'bambou noir'

Bailey

No locality recorded

Tufted bamboo to 7 m tall with stems flattened on one side and becoming blackish, with leaf blades to 10 cm long and 1.5 cm wide.

Rhynchelytrum Hochst.

R. repens (Willd.)C.E.Hubb. (*Tricholaena rosea*)

Stoddart (3)

Jef 499(S), Rob 2442(S)

Localities recorded: M,Den

Annual or perennial tufted grass to 1.2 m tall, with spreading panicle of silky feathery red spikelets, each to 6 mm long and falling early.

Rottboelia L.f.

R. exaltata L.f.

Summerhayes

Dup 3/31(K)

Localities recorded: M

Indigenous coarse annual to 3.5 m tall in wet places with solitary spikes in the axils of the upper leaves with a thickened axis bearing a pair of small spikelets in each joint, and breaking up when mature into sections 6 mm long. (Fig. 210)

Saccharum L.

S. officinarum L.   'canne à sucre, sugar cane'

Bailey, Gwynne, Stoddart (3), Baumer, Robertson (4),(5)

Localities recorded: M(sr only),F,N,Bir,Rem

Stout cultivated grass with stems to 6 m, long wide leaves and occasionally a

293

large white feathery panicle to 50 cm long. Sugar is produced by crystallizing the juice squeezed from the stems. Used medicinally in Seychelles.

S. spontaneum L. var. aegyptiacum Hack.

Dup 4/29(K)

Localities recorded: M

Ornamental grass, a perennial to 4.5 m tall with creeping rhizomes, leaf blades to 1.5 m long and feathery silvery white panicle to 60 cm long, each spikelet to 4 mm long with white hairs to 1.8 cm long.

Sacciolepis Nash.

S. curvata (L.)Chase

Jef 693(S), Sch 11709(K), Pro 4483(K), Rob 2402(S)

Localities recorded: M

Slender perennial straggling grass with a wide delicate panicle to 15 cm long of small green one sided spikelets each to 3 mm long, usually in shady places.

Sclerodactylon Stapf

S. macrostachyum (Benth.)A.Camus   'herbe piquant, herb bayonette'

Renvoize, Fosberg (2)

Localities recorded: Ald,Ass

Densely tufted glaucous perennial grass to 90 cm tall with inflorescence of usually 2 digitately arranged spikes, each to 12 cm long, crowded on one side with flattened long many-flowered spikelets to 2.5 cm long.

Setaria P.Beauv.

S. barbata (Lam.)Kunth

Summerhayes

Squ G121(K), Jef 714(S), Pro 4065,4115,4162(S), Bru/War 102(S), War 189(S), Rob 2444,2590,2642(S)

Localities recorded: M,S,F,N,A

Straggling grass rooting at the nodes and to 50 cm tall with wide 'pleated' leaf blades and a panicle of short racemes held at right angles to the rachis with small crowded ovate purplish spikelets, each to 2 mm long and short bristles to 6 mm long.

Sorghum Moench

S. arundinaceum (Desv.)Stapf (*S.verticilliflorum*)   'mille marron, wild sorghum'

Summerhayes

Pro 4454(K), Rob 2435(S)

Localities recorded: M,S,P

Stout tufted grass to 3 m with spreading panicle with hairy spikelets of two kinds, one almost round and pale becoming reddish brown, the other stalked, pointed and green.

S. bicolor (L.)Moench (*S. vulgare*) 'petit mille, sorghum, millet'

Bailey

No locality recorded

Vigorous grass to 6 m with tight or spreading inflorescences with fat sessile spikelets containing the edible grain and narrow stalked spikelets.

Sporobolus R.Br.

S. aldabrensis Renvoize

Renvoize, Fosberg (2)

Localities recorded: Ald

Endemic loosely tufted annual grass with wiry leaves and inflorescence a narrow spike to 14 cm long with spikelets to 1.5 mm long on short branches appressed to the main axis.

S. diander (Retz.)P. Beauv.

Pro 3964(S),4387(K)

Localities recorded: M,S

Slender tufted grass with long thin leaf blades and narrow panicle with short branches with minute grey green spikelets, each to 1.8 mm long.

S. pyramidalis P.Beauv.

Pro 4491(K)

Localities recorded: M

Erect tufted grass to 1.5 m with greyish green narrow panicle to 40 cm long with slender appressed branchlets closely covered with small ovate spikelets, each to 2 mm long.

S. tenuissimus (Schrank)Kuntze (*S. minutiflorus*)

Summerhayes

Jef 439(S),Squ G3(K)

Localities recorded: M,P,L

Slender annual grass to 80 cm with long leaf blades and delicate much branched oblong panicle with many minute grey green spikelets, each to 1 mm long, which gape at maturity to show the brown grain.

S. testudinum Renvoize

Renvoize, Fosberg (2)

Localities recorded: Ald

Endemic tufted annual or perennial with wiry leaves, a narrow spike, and spikelets to 1 mm long on appressed side branches, whole plant smaller than *S. aldabrensis.*

S. virginicus (L.)Kunth

Summerhayes, Sauer, Renvoize, Fosberg (2), Stoddart (2),(3), Robertson (7)

Squ G4(K), Ves 5399,5945,6147(K), Jef 1178(S), Fos 52032(S), Pro 3941,4150,4216(S), Bru/War 25(S), Rob 2740,3028(S)

Localities recorded: M,P,F,C,A,Co,Se,L,R,Bir,Den,Pla,Dar,Jos,Pie,Ald,Ass, Cos,Ast

Indigenous perennial rhizomatous grass at beach crest with stiff leaf blades to 10 cm long and a spike-like panicle to 10 cm long with silvery grey spikelets, each to 2.5 mm long, untidily arranged on short appressed branchlets. (Fig. 211)

Stenotaphrum Trin.

S. clavigerum Stapf

Renvoize, Fosberg (2)

Localities recorded: Ald,Ass

Endemic delicate tufted annual grass to 25 cm tall with slender stems, short leaves and spikes to 8 mm long with 5 spikelets, each to 2 mm long, sunk in depressions on opposite sides of the main axis.

S. dimidiatum (L.)Brongn. (*S. complanatum*)   'chiendent, herbe coco'

Summerhayes, Bailey, Gwynne, Piggott, Procter (2), Stoddart (3), Wilson, (4), Robertson (6), Renvoize

Squ G36(K), Jef 390(S), Sch 11776(K), Fos 52033(S), Pro 3946(S), 4390(K), Bru/War 121(S), Rob 2377,2396,2574,2679,2867,2894,3085(S),2510(K)

Localities recorded: M,S,P,F,N,C,A,An,L,Co,Bir,Coe,Des,Dar,Jos,Poi,Mar, Alp,Pro,Far

Indigenous stoloniferous mat-forming perennial grass with obtuse leaf blades to 12 cm long, and spike-like inflorescence with a broad flattened axis with short racemes of 3 to 5 spikelets, each to 5 mm long, sunk into one side. (Fig. 212)

S. micranthum (Desv.)C.E.Hubb. (*S. subulatum*)   'herbe la seine'

Summerhayes, Bailey, Sauer, Renvoize, Fosberg (2), Stoddart (1),(2),(3), Wilson (3)

Squ G1(K), Ves 5401,5806,5917,6148(K), Jef 1208(S), Fos 52099,52043(S), Pro 4041,4103(S), 4426(K), Fra 441(S), Bru/War 14(S), Rob 2311,2871, 2895,2971,3027,3134(S),2501(K)

Localities recorded: P,A,L,Co,Cu,Se,V,R,Bir,Den,Pla,Coe,Rem,Des,Dar,
Jos,Alp,Pie,Far,Cos,Ast,Def

Spreading or erect annual grass, rooting at the nodes, with short pointed leaf
blades and cylindrical spikes to 15 cm long with many spikelets, each to 3.5
mm long, arranged in short racemes sunk into depressions on opposite sides
of the main axis. Whole plant sometimes reddish.

Thysanolaena Nees

T. maxima O.Kuntze (*Panicum acariferum*)   'bambou de chine, tiger grass'

Bailey

Rob 2475(S)

Localities recorded: M

Stout ornamental grass to 3.5 m with broad leaves in two ranks and a
spreading drooping panicle with long branches and many spikelets, each to 5
mm long, with stems yellowish and midrib orange.

Tripsacum L.

T. laxum Nash   'guatemala grass'

Bailey

No locality recorded

Cultivated robust perennial grass with ♂ florets at the top of the spikelet and ♀
flowers at the base in cavities in the rachis. Used for fodder and mulch.

Urochloa P. Beauv.

U. paspaloides Presl

Pro 4526(K), Rob 2454,2563,2750(S)

Localities recorded: M,N

Slender straggling loosely tufted grass to 50 cm rooting at the nodes, with up to
4 spreading racemes to 10 cm long, with narrow pointed spikelets, each to 4
mm long and sometimes with a purplish tinge, arranged in 2 untidy ranks.

Vetiveria Bory

V. zizanioides (L.)Nash   'cuscus, vetiver'

Bailey, Lionnet (1), Baumer

Rob 2734(S)

Localities recorded: M,F

Cultivated densely tufted perennial wiry grass with long narrow leaf blades,
rough at the edge, to 75 cm long, usually sterile. Used for boundaries and
erosion control and to make mats, the roots yielding vetiver oil. Used
medicinally in Seychelles.

Zea L.

Z. mays L.   'mais, maize, sweet corn, corn'

Bailey, Fosberg (2), Stoddart (1),(3), Baumer

Localities recorded: M(sr only),Bir,Des,Far,Ald,Cos

Tall cultivated grass to 3 m with wide drooping leaf blades, ♂ racemes at the top of the plant (tassel) and ♀ spikes (up to 3) in the lower leaf axils with the styles protruding (silk) which when fertilized yield the cob with edible grains (white, yellow and sometimes red), each to 1.5 cm long. Used medicinally in Seychelles.

GYMNOSPERMS

ARAUCARIACEAE (207)

Agathis Salisb.

A. robusta (F.J.Muell.)F.M.Bailey   'queensland kauri pine'

Lionnet (2), (2a)

Localities recorded: M(BG)

Cultivated tall tree with opposite oblong juvenile leaves and elliptic adult leaves to 9 cm long, separate ♂ and ♀ flowers and cones to 12 cm long.

A. australis Hort. ex Lindl.

Lionnet (2a)

Localities recorded: M(BG)

Probably a misidentification as this species is rarely cultivated and specimens identified as such are usually *A. robusta*.

Araucaria Juss.

A. columnaris (G.Forst.)Hook.   'araucaria, monkey puzzle, new caledonian pine'

Bailey, Lionnet (2)

Localities recorded: M(BG)

Ornamental tree with whorled branches, the mature tree appearing columnar below an abruptly spreading short crown, the 'leaves' stiff, flat and overlapping, and cones to 12 cm long.

A. heterophylla (Salisb.)Franco   'araucaria, monkey puzzle, norfolk island pine'

Bailey, Robertson (7)

Localities recorded: M(sr only),Dar

Ornamental tree with whorled branches in a pyramidal shape, the branchlets erect, and stiff flat overlapping 'leaves'.

PINACEAE (208)

Pinus L.

P. caribaea Morelet   'caribbean pine, cuba pine'

Bailey

Localities recorded: M(sr only),P(sr only)

Cultivated timber tree with dark glossy green needles in clusters of 3 to 5, to 25 cm long, with oblong cones to 15 cm long.

CUPRESSACEAE (210)

Cupressus L.

C. pendula   'cypres'

Bailey

No locality recorded

There are 6 different species called *C.pendula*, one synonomous with *Thuja orientalis*, but as no specimen was seen it is not known which species Bailey meant.

Juniperus L.

J. procera Endl.   'juniper, african juniper'

Bailey

No locality recorded

Cultivated tree to 30 m with rounded crown, leaves scale-like or lanceolate and spreading with yellow catkin-like ♂ cones and brown or purplish ♀ cones.

CYCADACEAE (216)

Cycas L.

C. thuarsii Gaud. (*C. circinalis*)   'sagou palm, false sago palm'

Bailey, Robertson (7)

Localities recorded: P(sr only),Dar

Palm-like tree with stout trunk to 5 m and pinnate leaves to 2 m long crowded at the apex of the trunk, the petioles spiny and the leaflets dark glossy green to 30 cm long and 1 cm wide, ♂ plants with terminal reddish yellow cones of flowers, ♀ plants with rusty brown terminal carpellary leaves and ellipsoid fruit to 5 cm long.

ADDENDA

The main additions of source material are as follows.

Aver. & Kudr. Averyanov, L.V. & Kudriavtzeva, E.P. Contribution to a study of the Flora and Vegetation of Seychelles. V.L. Komarov Botanical Institute of the USSR Academy of Sciences. Leningrad. 1987.

Friedmann Friedmann, F. Flowers and Trees of Seychelles. Office de la Recherche Scientifique et Technique Outre-Mer. Paris. 1986.

Kew staff A number of staff at the Royal Botanic Gardens, Kew, have kindly looked through the galley proofs and made various suggestions concerning the current application of names. My particular thanks are offered to Mr Charles Jeffrey, Dr Roger Polhill, Mr Steve Renvoize and Dr Bernard Verdcourt, as well as the specialists for various families.

Milne Some additions are included from a collection of plants made by Mr Richard Milne from Mahe and Praslin in July 1988.

FLACOURTIACEAE (17/3)

p. 8 *Aphloia seychellensis* Hemsley is now considered to be a synonym of *A. theiformis* (Vahl) Benn. & Brown from Africa Kew staff.

PITTOSPORACEAE (18)

p. 10 *Pittosporum wrightii* Hemsley is now placed as a subspecies of *P. senacia*, as *P. senacia* Putt. subsp. *wrightii* (Hemsley) Cuf. Kew staff.

TILIACEAE (33/1)

p. 25 *Triumfetta suffruticosa* Bl. Aver. & Kudr. Locality not given. Shrubby herb to several m, with larger fruits than *T. rhomboidea*, 2–3 cm across.

BALSAMINACEAE (38/8)

p. 30 *Impatiens thomassetii* Hook.f. is now regarded as a synonym of *I. gordonii*. Kew staff.

VITACEAE (50/1)

p. 44 *Cissus rotundifolia* Vahl Friedmann Locality: M Strong climber with fleshy round branches, large waxy-fleshy round leaves, small green flowers and red berries.

LEGUMINOSAE (57)

p. 50 *Cassia* is now generally treated as three separate genera. *C. fistula, C. grandis* and *C. javanica* L. (*C. nodosa* Roxb.) remain in *Cassia*. *C. mimosoides* becomes *Chamaecrista mimosoides* (L.) Green and *C. aldabrensis* should be referred to the same genus, but no combination has been made yet. *C. alata* = *Senna alata* (L.) Roxb.; *C. hirsuta* = *Senna hirsuta* (L.) Irwin & Barneby; *C. multijuga* = *Senna multijuga* (Rich.) Irwin & Barneby; *C. occidentalis* = *Senna occidentalis* (L.) Link; *C. polyphylla* = *Senna polyphylla* (Jacq.) Irwin & Barneby; *C. sophera* = *Senna sophera* (L.) Roxb. and *C. tora* = *Senna tora* (L.) Roxb.   Kew staff.

p. 58 *Albizia falcata* is now referred to *Paraserianthes* as *P. falcataria* (L.) Nielsen (*Albizia falcataria*).   Kew staff.

p. 59 *Dichrostachys microcephala* Renv. has now been treated as a variety, becoming *Gagnebina commersoniana* (Baill.) R. Vig. var. *aldabrensis* Lewis & Guinet   Kew staff.

p. 59 The correct name for *Entada pursaetha* is now *E. rheedii* Sprengel, but the circumscription is unchanged   Kew staff.

p. 61 *Samanea saman* is now referred to *Albizia* as *A. saman* (Jacq.) F. Muell.   Kew staff.

p. 61 **Aeschynomene americana** L. Milne 47   Locality: M   Erect or spreading herb to 1 m or more, with sparse bristly hairs, pinnate leaves with 20–60 narrow, sensitive, several-nerved, slightly curved leaflets to 1.5 cm long, with generally purplish mauve flowers to 8 mm long, and a 3–9-jointed fruit to 3 cm long, the articles rounded on the lower side.

MYRTACEAE (67/1)

p. 86 *Eugenia caryophyllus,* the clove, is now generally referred to *Syzygium* as *S. aromaticum* (L.) Merr. & Perry   Kew staff.

MELASTOMATACEAE (68)

p. 91 **Clidemia hirta** (L.) D.Don Aver. & Kudr.   Locality: S   Shrub 1–3 m tall, a weed, with pairs of slightly unequal sized hairy ovate leaves, 3-nerved from base, and short sprays of white to mauve flowers 2 cm across, the stamens equal in length, fruit a berry within the persistent bristly calyx.

CUCURBITACEAE (75)

p. 98 **Benincasa hispida** (Thunb.) Cogn. 'wax gourd'   Kew staff, Jeffrey   Locality: M (sr)   Cultivated hispid trailing annual herb with branched tendrils, palmately lobed leaves, yellow flowers and large ellipsoid indehiscent fruits with a thick waxy bloom at maturity.

ARALIACEAE (81)

p.108 *Geopanax procumbens* Hemsley is now referred to *Schefflera,* as *S. procumbens* (Hemsley) Frodin   Kew Staff.

p.108   *Indokingia crassa* is now referred to *Gastonia* as *G. crassa* (Hemsley) Friedm.

RUBIACEAE (84/1)

p.110   *Canthium bibracteatum* (Baker) Hiern is now referred to *Pyrostria bibracteata* (Baker) Cavaco   Kew staff.

p.112   *Hedyotis corallicola* Fosberg and *H. prolifera* Fosberg should be transferred to *Oldenlandia*, but no combinations are available yet. *H. lancifolia* var. *brevipes* can be referred back to *Oldenlandia*, as *O. lancifolia* (Schum.) DC. var. *brevipes* Brem.   Kew Staff.

p.118   *Randia lancifolia* is now referred to *Paragenipa lancifolia* (Boj. ex Baker) Tirv. & Robbr.   Kew Staff.

p.118   *Randia sericea* is now referred to *Glionnetia sericea* (Bak.) Tirv.   Kew staff.

COMPOSITAE (88)

p.119   **Spermacoce sp.**, possibly *S. assurgens* Ruiz & Pavon (*S. laevis*) Milne 49 Locality: M   Annual weed to 50 cm, with opposite elliptic to ovate leaves to 5 cm long, clusters of small white flowers and capsules to 4 mm long, with a single seed in each of the two chambers.

p.122   The two species of *Chrysanthemum* are now referred to *Dendranthema*, as *D. indicum* (L.) Des Moul. (*Chrysanthemum indicum* L.) and *D.* × *grandiflorum* (Ramat.) Kitam. (*Chrysanthemum sinense* Sabine)   Kew staff.

p.124   *Eupatorium ayapana* Vent. now placed in *Ayapana*, as *A. triplinervis* (Vahl) R. King & H. Robinson   Kew staff.

p.128   *Melanthera biflora* (L.) Wild is now referred to *Wollastonia* as *W. biflora* (L.) DC.   Kew staff.

p.129   *Spilanthes mauritiana* (Rich. ex Pers.) DC. is now referred to *Acmella*, as *A. caulirhiza* Del.   Kew staff.

p.130   *Vernonia grandis* (DC.) J. Humb. is now treated as a subspecies of *V. colorata*, as *V. colorata* (Willd.) Drake subsp. *grandis* (DC.) C. Jeffrey Kew staff.

p.130   **Wedelia trilobata** (L.) Hitchc. Aver. & Kudr.   Locality: M   Stems creeping, with 3-lobed leaves to 4 cm long, flower heads yellow, to 2.5 cm across, with spreading ray florets.

BORAGINACEAE (112/1)

p.150   *Tournefortia argentea* L.f. is now referred to *Argusia*, as *A. argentea* (L.f.) Heine   Kew staff.

SCROPHULARIACEAE (115)

p.163   **Otacanthus coeruleus** Lindl. Kew staff, Jeffrey   Locality: M (sr) Perennial herb to 80 cm, with opposite elliptic serrate leaves, conspicuous tubular 2-lipped flowers 50 mm long in clusters at the tops of the stems, and conical capsules 7 mm long.

ACANTHACEAE (122)

p.172   **Pseuderanthemum sp. aff. tunicatum** (Afzel.) Milne-Redh. Friedmann
Locality: S   Apparently a local endemic.

p.172   **Ruellia tuberosa** L. Milne 16   Locality: P   Slightly woody herb to 1 m,
with opposite elliptic leaves to 15 cm, inflorescences of several mauve to
reddish purple flowers with spreading lobes, to 4 cm across, and spindle-
shaped capsules to 3 cm long.

VERBENACEAE (125/1)

p.173   *Clerodendrum fragrans* (Vent.) Willd. should be more correctly called *C.
phillippinum* Schauer   Kew staff.

p.174   **Clerodendrum ✕ speciosum** Carrière Friedmann   Locality not cited.
A hybrid between *C. splendens* and *C. thomsonae*, differing from the latter
(see p. 174) by the bright red flowers with a purple calyx.

NYCTAGINACEAE (128)

p.186   **Pisonia sechellarum** Friedm. has recently been described from
Silhouette and may also have occurred formerly on Mahe (summit of
Morne Blanc). Endemic tree to 15 m, elliptic leaves to 30 cm mostly in
whorls, clusters of small pale yellow flowers, each 4–5 mm long,
arranged into a panicle and fruits similar to *P. grandis* but up to 4 cm
long.

p.186   **Pisonia umbellifera** (J.R. & G. Forst.) Seem. or related species. 'mapou
de grand bois'   Friedmann   Locality: S   Tree about 10–15 m, trunk
attaining 1 m diameter, soft wood saturated with water.

AMARANTHACEAE (130)

p.189   **Amaranthus lividus** L. Aver. & Kudr.   Locality: Poi   Glabrous annual
to 1 m, similar to *A. viridis*, but capsules smooth and compressed.

PIPERACEAE (139/1)

p.195   **Piper sp.** 'wild pepper' Friedmann   Locality: S   Abundant creeper
on trees.

p.195   *Piper umbellatum* L. now usually referred to *Pothomorphe*, as *P. umbellata*
(L.) Miq.   Kew staff.

EUPHORBIACEAE (151/1)

p.202   **Euphorbia peplus** L. Aver. & Kudr.   Locality: M   Glabrous annual
herb, a weed, with obovate leaves to 25 cm, and trichotomous leafy
inflorescences bearing tiny flowers and 3-lobed capsules with fleshy
ridges along the angles.

p.206   *Jatropha pandurifolia* Andr. is now referred to *J. integerrima* Jacq.   Kew
staff.

p.208 *Phyllanthus longifolius* Lam. Aver. & Kudr. Locality: Ass Probably a misidentification, possibly of *P. maderaspatensis* (see p. 205). *P. revaughanii* Coode (*P. longifolius*) is endemic to Mauritius.

p.210 *Securinega virosa* (Roxb. ex Willd.) Baill. is now referred to *Flueggea*, as *F. virosa* (Roxb. ex Willd.) Voigt Kew staff.

URTICACEAE (153/1)

p.211 *Laportea interrupta* (L.) Chew Aver. & Kudr. Localities: S, Poi Similar to *L. aestuans* but with a long interrupted spike-like inflorescence.

p.211 *Obetia ficifolia* Gaud. is now considered not to include the plants on Aldabra, which have been referred to a new endemic species *Obetia aldabrensis* Friis Kew staff.

p.212 *Procris latifolia* Blume may be better referred to a number of species, of which *P. insularis* H. Schroter is endemic to Seychelles Kew staff.

MORACEAE (153/5)

p.219 *Chlorophora excelsa* (Welw.) Benth. is now referred to *Milicia*, as *M. excelsa* (Welw.) C.C. Berg Kew staff.

ORCHIDACEAE (169)

p.220 **Bulbophyllum sp.** Friedmann Locality: presumably M Larger than *B. intertextum*, with white flowers, as yet unidentified.

p.221 *Cirrhopetalum umbellatum* is now referred to *Bulbophyllum* as *B. longiflorum* Thouars Kew staff.

p.221 **Cynorkis seychellarum** Aver. Aver. & Kudr. Localities: M, S, P, Fe A newly described species; the type is from Praslin.

p.222 **Eulophia pulchra** (Thouars) Lindl. (*Eulophidium pulchrum*) Friedmann Localities: M, S, P A large and vigorous terrestrial orchid.

p.224 **Oeoniella aphrodite** Schltr. Friedmann erroneously refers to this species as *O. polystachys* Schltr., a much larger plant not known from Seychelles.

HYPOXIDACEAE (174/2)

p.235 *Curculigo rhizophylla* (Baker) Dur. & Schinz and *C. sp.* are now referred to a new endemic genus *Hypoxidia*, as *H. rhizophylla* (Baker) Friedm. and *H. maheensis* Friedm. respectively. *H. maheensis* differs from *H. rhizophylla* most conspicuously by its pale yellow flowers flushed with pink and its broader leaves, which do not root at the tips.

ARACEAE (191)

p.258 **Epipremnum pinnatum** (L.) Engl. cv. Aureum (*E. aureum*) Friedmann Locality: P Vigorous ornamental climber, mature leaves irregularly dissected, green blotched yellow, up to 75 cm long, rarely flowering.

LEMNACEAE (192)

p.259 **Lemna perpusilla** Torrey Aver. & Kudr. Locality: Fe Probably same as *L. sp.* (p. 259).

CYPERACEAE (199)

p.264 **Cyperus cuspidatus** Kunth Aver. & Kudr. Locality: D Slender annual to 15 cm, with 4–20 reddish brown spikelets, each 4–10 mm long, in clusters 1–2 cm wide.

p.265 **Cyperus iria** L. Aver. & Kudr. Localities: D, Fe Tussocky annual to 60 cm, spikes irregular in shape and length, spikelets golden to yellowish green, 2–10 mm long.

p.269 **Kyllinga tenuifolia** Steud. (*K. triceps*) Aver. & Kudr. Localities: M, S, Fe Tufted perennial to 30 cm, swollen bases of stems covered with brown scales and old leaf sheaths, leaves flat or incurved, often with dark purple markings, inflorescence a pale green to greyish white head 4–9 mm long.

GRAMINEAE (200)

p.292 **Pennisetum glaucum** (L.) R.Br. (*Setaria glauca*) 'bulrush millet' Aver. & Kudr. Localities: M, S, D, Far Widely cultivated crop plant.

p.298 **Zoysia matrella** (L.) Merr. (*Z. pungens*) Aver. & Kudr. Localities: Poi, Far, Ass Small, slender, fine leaved species, introduced as a lawn grass.

Index to vernacular names

* Also generic name

Index to family and generic names

Lightning Source UK Ltd.
Milton Keynes UK
29 October 2009

145515UK00001B/48/A